Pre-Calculus Workbook

FOR

DUMMIES®

**by Michelle Rose Gilman,
Christopher Burger,
Karina Neal**

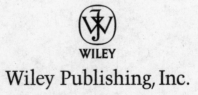

WILEY

Wiley Publishing, Inc.

Pre-Calculus Workbook For Dummies®

Published by
Wiley Publishing, Inc.
111 River St.
Hoboken, NJ 07030-5774
www.wiley.com

Copyright © 2009 by Wiley Publishing, Inc., Indianapolis, Indiana

Published by Wiley Publishing, Inc., Indianapolis, Indiana

Published simultaneously in Canada

For general information on our other products and services, please contact our Customer Care Department within the U.S. at 877-762-2974, outside the U.S. at 317-572-3993, or fax 317-572-4002.

For technical support, please visit www.wiley.com/techsupport.

Wiley also publishes its books in a variety of electronic formats. Some content that appears in print may not be available in electronic books.

Library of Congress Control Number: 2009923971

ISBN: 978-0-470-42131-4

Manufactured in the United States of America

10 9 8 7 6 5 4 3 2 1

WILEY

About the Authors

Michelle Rose Gilman is proud to be known as Noah's mom (Hi, Noah!). A graduate of the University of South Florida, Michelle found her niche early — at 19 she was already working with emotionally disturbed and learning-disabled students in hospital settings. At 21, she made the trek to California. There she discovered her passion for helping teenage students become more successful in school and life. What started as a small tutoring business in the garage of her California home quickly expanded and grew to the point where traffic control was necessary on her residential street.

Today, Michelle is the founder and CEO of the Fusion Learning Center/Fusion Academy, a private school and tutoring/test-prep facility in Solana Beach, California, serving more than 2,000 students per year. She has taught tens of thousands of students since 1988. In her spare time, Michelle created the Mentoring Approach to Learning and authored *The ACT For Dummies, Pre-Calculus For Dummies, AP Biology For Dummies, AP Chemistry For Dummies, Chemistry Workbook For Dummies,* and *The GRE For Dummies.* She currently specializes in motivating the unmotivated adolescent, comforting shell-shocked parents, and assisting her staff of 27 teachers.

Michelle lives by the following motto:

"There are people content with longing; I am not one of them."

Christopher Burger graduated with a Bachelor of Arts degree in mathematics from Coker College in Hartsville, South Carolina, with minors in art and theater. He has taught math for more than 10 years and has tutored subjects ranging from basic math to calculus for 20 years. He is currently the Director of Independent Studies for Fusion Learning Center and Fusion Academy in Solana Beach, California, where he not only teaches students one-on-one but also writes curriculum, oversees a staff of 27 teachers, and maintains a high level of academic rigor within the school. When not at school, Christopher can be found in local theaters directing, acting, stage managing, or doing pretty much any job that they'll let him do. Chris is also one of the authors of *Pre-Calculus For Dummies.*

Karina Neal graduated with a Bachelor of Science degree in combined sciences with an emphasis in psychology from Santa Clara University. Additionally, she received her certificates in educational therapy and in college counseling from the University of California, San Diego. From an early age, teaching and tutoring have been her passions — from starting her own tutoring business in high school to helping found the Fusion Learning Center and Fusion Academy in Solana Beach, California. As that institution's Director of Tutoring and Mentoring, Karina teaches all levels of mathematics and science, provides special education and college counseling consultation, and oversees a staff of 27 tutors and teachers. Karina has over 18 years of experience in the education field and continues to tutor and teach students in a wide range of subjects, from remedial writing to calculus. Besides being a closet math and science geek, Karina is dedicated to the success of her students and believes that all students can learn.

Dedication

We would like to dedicate this book to every student we've ever taught — each one of you taught us something in return. Also, to our families and friends who supported us during the writing.

Authors' Acknowledgments

We would like to acknowledge Bill Gladstone, our wonderful agent from Waterside, for the opportunity to write this book; Nicholas Angelo for being the scanning king; Natalie Harris, project editor extraordinaire; copy editor Todd Lothery; technical editor David Herzog; acquisitions editors Tracy Boggier and Erin Mooney, who, for unknown reasons, continue to want to work with us; and to everyone who has lent a helping hand or eye or brain, we couldn't do it without you: TL, VS, BN, NG, and Clyde.

Publisher's Acknowledgments

We're proud of this book; please send us your comments through our Dummies online registration form located at http://dummies.custhelp.com. For other comments, please contact our Customer Care Department within the U.S. at 877-762-2974, outside the U.S. at 317-572-3993, or fax 317-572-4002.

Some of the people who helped bring this book to market include the following:

Acquisitions, Editorial, and Media Development

Project Editor: Natalie Faye Harris

Acquisitions Editor: Tracy Boggier

Copy Editor: Todd Lothery

Assistant Editor: Erin Calligan Mooney

Editorial Program Coordinator: Joe Niesen

Technical Editor: David Herzog

Editorial Manager: Christine Meloy Beck

Editorial Assistants: Jennette ElNaggar, David Lutton

Cartoons: Rich Tennant (www.the5thwave.com)

Composition Services

Project Coordinator: Patrick Redmond

Layout and Graphics: Carl Byers, Carrie A. Cesavice, Reuben W. Davis, Nikki Gately

Proofreaders: John Greenough, Leeann Harney

Indexer: Broccoli Information Mgt.

Special Help

Peter Mikulecky

Publishing and Editorial for Consumer Dummies

> **Diane Graves Steele,** Vice President and Publisher, Consumer Dummies

> **Kristin Ferguson-Wagstaffe,** Product Development Director, Consumer Dummies

> **Ensley Eikenburg,** Associate Publisher, Travel

> **Kelly Regan,** Editorial Director, Travel

Publishing for Technology Dummies

> **Andy Cummings,** Vice President and Publisher, Dummies Technology/General User

Composition Services

> **Gerry Fahey,** Vice President of Production Services

> **Debbie Stailey,** Director of Composition Services

Contents at a Glance

Table of Contents

Introduction

• •

You've just picked up the best workbook ever to help you with pre-calculus, if we do say so ourselves. If you've gotten this far in your math career, congratulations! Many students choose to stop their math education after they complete Algebra II, but not you!

If you've picked up this book (obviously you have; you're reading this sentence, duh!), maybe some of the concepts in pre-calc are giving you a hard time, or perhaps you just want more practice. Maybe you're deciding whether you even want to take pre-calc at all. This book fits the bill for all those reasons. And we're here to cheerlead you on during your pre-calc adventure. Look, if you've gotten this far in math, you're no "Dummie," so don't let the title throw you!

We know that you'll find this workbook chock-full of valuable practice problems and explanations. In instances where you feel you may need a more thorough explanation, please refer to *Pre-Calculus For Dummies* (we wrote that one too — yes, we are math geeks). In some areas of the book, we even refer you to *Pre-Calculus For Dummies* ourselves. We set up this workbook to directly coincide with the format of *Pre-Calculus For Dummies* in an effort to make it really easy for you to use the two together, if you wish. This book, however, is a great stand-alone workbook if you need extra practice, need just a brushup in certain areas, or just can't stand our jokes in the other book.

About This Book

Don't let pre-calc scare you. When you realize that you already know a whole bunch from Algebra I and Algebra II, you'll see that pre-calculus is really just using that old information in a new way. And even if you're scared, we're here with you, so no need to panic. Before you get ready to start this new adventure, you need to know a few things about this book.

This book isn't a novel. It's not meant to be read in order from beginning to end. You can read any topic at any time, but we've structured it in such a way that it follows the "normal" curriculum. This is hard to do because most states don't have state standards for what makes pre-calc pre-calc. We looked at a bunch of curriculums, though, and came up with what we think is a good representation of a Pre-Calc course. Sometimes, we may include a reference to material in another chapter, and we may send you there for more information.

Instead of placing this book on a shelf and never looking at it again, or using it as a doorstop (thanks for the advertisement, in either case), we suggest you follow one of two alternatives:

✔ Look up what you need to know when you need to know it. The index, table of contents, and even the contents at a glance section will all direct you where to look.

✔ Start at the beginning and read through. This way, you may be reminded of an old topic that you had forgotten (anything to get those math wheels churning inside your head). Besides, practice makes perfect, and the problems in this book are a great representation of the problems found in pre-calc textbooks.

Conventions Used in This Book

For consistency and ease of navigation, this book uses the following conventions:

- ✔ Math terms are *italicized* when they're introduced or defined in the text.
- ✔ Variables are *italicized* to set them apart from letters.
- ✔ The symbol for imaginary numbers is a lowercase *i*.

Foolish Assumptions

We don't assume that you love math the way we do as professional math geeks. We do assume, though, that you picked this book up for a reason of your own. Maybe you want a preview of the course before you take it, or perhaps you need a refresher on the topics in the course, or maybe your kid is taking the course and you're trying to help him be successful.

Whatever your reason, we assume that you've encountered most of the topics in this book before, because for the most part, the topics are reviews of ones you've seen in algebra or geometry.

How This Book Is Organized

This book is divided into five parts dealing with the most commonly taught topics of pre-calc.

Part I: Foundation (And We Don't Mean Makeup!)

The chapters in Part I begin at the beginning. First we review basic material from Algebra II. We then cover real numbers and what you'll be asked to do with them. Next up are functions of all kinds (polynomials, rational, exponential, and logarithmic): graphing them and performing operations with them.

Part II: Trig Is the Key: Basic Review, the Unit Circle, and Graphs

The chapters in Part II review trig ratios and word problems for trig. Then we show you how to build the unit circle, how to solve trig equations, and how to graph trig functions. Some of these topics may be review for you as well; that really depends on how much trig was covered in your Algebra II course.

Part III: Advanced Trig: Identities, Theorems, and Applications

The chapters in Part III cover basic and advanced identities. We cover the tricky trig proofs in this part. If you're asked to do trig proofs in your Pre-Calc course, you definitely want to check out our tips on how to handle them like a pro. We also cover some trig applications that can be solved using the Law of Sines or the Law of Cosines.

Part IV: And the Rest . . .

The chapters in Part IV cover the topics from the remainder of the Pre-Calc course. We introduce complex numbers and how to work with them, and we explain conic sections and how to graph them. Because systems of equations tend to get harder in pre-calc, we begin with a review and build up to the tougher topics. Your Pre-Calc course may only focus on a couple of these topics, so be sure to pay attention to the table of contents here. Next, we move into sequences and series and introduce the binomial theorem, which helps you raise binomials to high powers. Last, we introduce the first topics of a calc course. Sometimes, these are the last topics you'll see in pre-calc, so we want to be sure to go over them.

Part V: The Part of Tens

This book has two handy lists at the end. The first list includes ten parent graphs: how to recognize them, how to graph them, and how to transform them. The second list covers common mistakes we often see that we'd like to help you avoid.

Icons Used in This Book

Throughout this book you'll see icons in the margins to draw your attention to something important that you need to know.

Pre-calc rules are exactly what they say they are — the rules of pre-calculus. Theorems, laws, and properties all make Pre-Calc an ironclad course — they must be followed at all times.

Tips are great, especially if you wait tables for a living! These tips are designed to make your life easier, which are the best tips of all!

The Remember icon is used one way: It asks you to remember old material from a previous math course.

 Warnings are big red flags that draw your attention to common mistakes you may get tripped up on.

Where to Go from Here

Pick a starting point in the book and go practice the problems there. If you'd like to review the basics first, start at Chapter 1. If you feel comfy enough with your algebra skills, you may want to skip that chapter and head over to Chapter 2. Most of the topics there are reviews of Algebra II material, but don't skip over something because you think you've got it under control. You'll also find in pre-calc that the level of difficulty in some of these topics gets turned up a notch or two. Go ahead — dive in and enjoy the world of pre-calc!

Part I
Foundation (And We Don't Mean Makeup!)

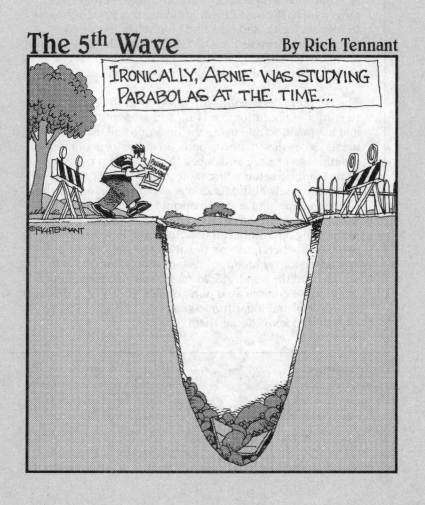

The 5th Wave   By Rich Tennant

IRONICALLY, ARNIE WAS STUDYING PARABOLAS AT THE TIME...

In this part . . .

Pre-calculus is really just another stop on the road to calculus. You started with the village of Algebra I, moved on to the small town of Geometry, made your way to Algebra II city, and now find yourself in the mega-metropolis known as Pre-Calculus. The skills, for the most part, are the same. This part takes those skills and reviews them (and, in some cases, expands on them).

The chapters here begin with a review of the basics: using the order of operations, solving and graphing equations and inequalities, and using the distance and midpoint formulas. Some new material pops up in the form of interval notation, so be sure and check that out. Then we move on to real numbers, including radicals. Everything you ever wanted to know about functions is covered in one of the chapters: graphing and transforming parent graphs, rational functions, and piece-wise functions. We also go over performing operations on functions and how to find the inverse. We then move on to solving higher degree polynomials using techniques like factoring, completing the square, and the quadratic formula. You also learn how to graph these complicated polynomials. Lastly, you discover exponential and logarithmic functions and what you're expected to know about them.

Chapter 1

Beginning at the Very Beginning: Pre-Pre-Calculus

· ·

In This Chapter

▶ Brushing up on order of operations

▶ Solving equalities

▶ Graphing equalities and inequalities

▶ Finding distance, midpoint, and slope

· ·

Pre-calculus is the stepping stone for Calculus. It's the final hurdle after all those years of math: Pre-algebra, Algebra, Geometry, and Algebra II. Now all you need is Pre-calculus to get to that ultimate goal — Calculus. And as you may recall from your Algebra II class, you were subjected to much of the same material you saw in Algebra and even Pre-algebra (just a couple steps up in terms of complexity — but really the same stuff). As the stepping stone, pre-calculus begins with certain concepts that you're expected to solidly understand.

Therefore, we're starting here, at the very beginning, reviewing those concepts. If you feel you're already an expert at everything algebra, feel free to skip past this chapter and get the full swing of pre-calc going. If, however, you need to review, then read on.

If you don't remember some of the concepts we discuss in this chapter, or even in this book, you can pick up another *For Dummies* math book for review. The fundamentals are important. That's why they're called fundamentals. Take the time now to review — it will save you countless hours of frustration in the future!

Reviewing Order of Operations: The Fun in Fundamentals

You can't put on your sock after you put on your shoe, can you? The same concept applies to mathematical operations. There's a specific order to which operation you perform first, second, third, and so on. At this point, it should be second nature. However, because the concept is so important as we continue into more complex calculations, we review it here.

Please excuse who? Oh, yeah, you remember this one — my dear Aunt Sally! The old mnemonic still stands, even as you get into more complicated problems. Please Excuse My Dear Aunt Sally is a mnemonic for the acronym PEMDAS, which stands for:

- ✔ **P**arentheses (including absolute value, brackets, and radicals)
- ✔ **E**xponents
- ✔ **M**ultiplication and **D**ivision (from left to right)
- ✔ **A**ddition and **S**ubtraction (from left to right)

The order in which you solve algebraic problems is very important. Always work what's in the parentheses first, then move on to the exponents, followed by the multiplication and division (from left to right), and finally, the addition and subtraction (from left to right). Because we're reviewing fundamentals, now is also a good time to do a quick review of properties of equality.

When simplifying expressions, it's helpful to recall the properties of numbers:

- ✔ **Reflexive property:** $a = a$. For example, $4 = 4$.
- ✔ **Symmetric property: If $a = b$, then $b = a$.** For example, if $2 + 8 = 10$, then $10 = 2 + 8$.
- ✔ **Transitive property: If $a = b$ and $b = c$, then $a = c$.** For example, if $2 + 8 = 10$ and $10 = 5 \cdot 2$, then $2 + 8 = 5 \cdot 2$.
- ✔ **Commutative property of addition (and of multiplication):** $a + b = b + a$. For example, $3 + 4 = 4 + 3$.
- ✔ **Commutative property of multiplication:** $a \cdot b = b \cdot a$. For example, $3 \cdot 4 = 4 \cdot 3$.
- ✔ **Associative property of addition (and of multiplication):** $a + (b + c) = (a + b) + c$. For example, $3 + (4 + 5) = (3 + 4) + 5$.
- ✔ **Associative property of multiplication:** $a \cdot (b \cdot c) = (a \cdot b) \cdot c$. For example, $3 \cdot (4 \cdot 5) = (3 \cdot 4) \cdot 5$.
- ✔ **Additive identity:** $a + 0 = a$. For example, $4 + 0 = 4$.
- ✔ **Multiplicative identity:** $a \cdot 1 = a$. For example, $-18 \cdot 1 = -18$.
- ✔ **Additive inverse property:** $a + (-a) = 0$. For example, $5 + -5 = 0$.
- ✔ **Multiplicative inverse property:** $a \cdot \dfrac{1}{a} = 1$. For example, $-2 \cdot (-\frac{1}{2}) = 1$.
- ✔ **Distributive property:** $a(b + c) = a \cdot b + a \cdot c$. For example, $5(3 + 4) = 5 \cdot 3 + 5 \cdot 4$.
- ✔ **Multiplicative property of zero:** $a \cdot 0 = 0$. For example, $4 \cdot 0 = 0$.
- ✔ **Zero product property: If $a \cdot b = 0$, then $a = 0$ or $b = 0$.** For example, if $x(2x - 3) = 0$, then $x = 0$ or $2x - 3 = 0$.

Q. Simplify: $\dfrac{6^2 - 4\left(3 - \sqrt{20+5}\right)^2}{|4-8|}$.

A. **The answer is 5.** Following our rules of order of operations, simplify everything in parentheses first.

Radicals and absolute value marks act like parentheses. Therefore, if any of the operations are under radicals or within absolute value marks, do those first before simplifying the radicals or taking the absolute value.

Simplify the parentheses by taking the square root of 25 and the absolute value

of –4: $\dfrac{6^2 - 4\left(3 - \sqrt{25}\right)^2}{|-4|} = \dfrac{6^2 - 4\left(3 - 5\right)^2}{4} =$

$\dfrac{6^2 - 4\left(-2\right)^2}{4}$. Now that the parentheses are simplified, you can deal with the exponents.

Square the 6 and the –2: $= \dfrac{36 - 4(4)}{4}$.

Although they're not written, parentheses are implied around the terms above and below a fraction bar. In other words, the expression $\dfrac{36 - 4(4)}{4}$ can also be written as

$\dfrac{\left[36 - 4(4)\right]}{4}$. Therefore, you must simplify the numerator and denominator before dividing the terms following the order of operations:

$= \dfrac{36 - 4(4)}{4} = \dfrac{36 - 16}{4} = \dfrac{20}{4} = 5$.

Q. Simplify: $\dfrac{\left(\dfrac{1}{8} + \dfrac{1}{3}\right) + \dfrac{3}{8}}{\dfrac{3}{18} + \dfrac{1}{9}}$.

A. **The answer is 3.** Using the associative property of addition, rewrite the expression to make the fractions easier to add:

$\dfrac{\left(\dfrac{1}{8} + \dfrac{3}{8}\right) + \dfrac{1}{3}}{\dfrac{3}{18} + \dfrac{1}{9}}$. Add the fractions with

common denominators, $\dfrac{\dfrac{4}{8} + \dfrac{1}{3}}{\dfrac{3}{18} + \dfrac{1}{9}}$, and

reduce the resulting fraction: $\dfrac{\dfrac{1}{2} + \dfrac{1}{3}}{\dfrac{3}{18} + \dfrac{1}{9}}$. Next,

find a common denominator for the fractions in the numerator and denominator:

$\dfrac{\dfrac{3}{6} + \dfrac{2}{6}}{\dfrac{3}{18} + \dfrac{2}{18}}$. Add these: $\dfrac{\dfrac{5}{6}}{\dfrac{5}{18}}$. Recognizing that

this expression is a division problem,

$\dfrac{5}{6} \div \dfrac{5}{18}$, multiply by the inverse and

simplify: $\dfrac{5}{6} \cdot \dfrac{18}{5} = \dfrac{5 \cdot 18}{6 \cdot 5} = \dfrac{\cancel{5} \cdot 18}{6 \cdot \cancel{5}} = \dfrac{3}{1} = 3$.

1. Simplify: $\dfrac{3\sqrt{\left(4-6\right)^2+\left(2-\left(-1\right)\right)^2}}{\left|-3-\left(-1\right)\right|}$.

Solve It

2. Simplify: $\dfrac{\left|-3\right|-\left|2\right|+\left(-1\right)}{\left|-7+2\right|}$.

Solve It

3. Simplify: $(2^3-3^2)^4(-5)$.

Solve It

4. Simplify: $\dfrac{\left|5\left|1-4\right|+6\right|}{3\left(-\dfrac{1}{6}+\dfrac{1}{3}\right)-\dfrac{1}{2}}$.

Solve It

Keeping Your Balance While Solving Equalities

Just as simplifying expressions is the basics of pre-algebra, solving for variables is the basics of algebra. Both are essential to more complex concepts in pre-calculus. Solving basic algebraic equations should be easy for you; however, it's so fundamental to pre-calculus, we give you a brief review here.

Solving linear equations with the general format of $ax + b = c$, where a, b, and c are constants, is relatively easy using properties of numbers. The goal, of course, is to isolate the variable, x.

One type of equation you can't forget is absolute value equations. The *absolute value* is defined as the distance from 0. In other words, $\left|x\right| = \begin{cases} x, x \geq 0 \\ -x, x < 0 \end{cases}$. As such, an absolute value has two possible solutions: one where the quantity inside the absolute value bars is positive and another where it's negative. To solve these equations, it's important to isolate the absolute value term and then set the quantity to the positive and negative values.

Q. Solve for x: $3(2x - 4) = x - 2(-2x + 3)$.

A. $x = 6$. First, using the distributive property, distribute the 3 and the -2: $6x - 12 = x + 4x - 6$. Combine like terms and solve using algebra: $6x - 12 = 5x - 6$; $x - 12 = -6$; $x = 6$.

Q. Solve for x: $|x - 3| + (-16) = -12$.

A. $x = 7, -1$. First, isolate the absolute value: $|x - 3| = 4$. Next, set the quantity inside the absolute value bars to the positive solution: $x - 3 = 4$. Then, set the quantity inside the absolute value bars to the negative solution: $-(x - 3) = 4$. Solve both equations to find two possible solutions: $x - 3 = 4$, $x = 7$; and $-(x - 3) = 4$, $x - 3 = -4$, $x = -1$.

5. Solve: $3 - 6[2 - 4x(x + 3)] = 3x(8x + 12) + 27$.

Solve It

6. Solve $\dfrac{x}{2} + \dfrac{x - 2}{4} = \dfrac{x + 4}{2}$.

Solve It

7. Solve $|x - 3| + |3x + 2| = 4$.

Solve It

8. Solve $3 - 4(2 - 3x) = 2(6x + 2)$.

Solve It

9. Solve $2|x - 3| + 12 = 6$.

Solve It

10. Solve $3(2x + 5) + 10 = 2(x + 10) + 4x + 5$.

Solve It

A Picture Is Worth a Thousand Words: Graphing Equalities and Inequalities

Graphs are visual representations of mathematical equations. In pre-calculus, you'll be introduced to many new mathematical equations and then be expected to graph them. We give you lots of practice graphing these equations when we cover the more complex equations. In the meantime, it's important to practice the basics: graphing linear equalities and inequalities.

These graphs are graphed on the *Cartesian coordinate system*. This system is made up of two axes: the horizontal, or *x*-axis, and the vertical, or *y*-axis. Each point on the coordinate plane is called a *Cartesian coordinate pair* (x, y). A set of these ordered pairs that can be graphed on a coordinate plane is called a *relation*. The *x* values of a relation are its *domain*, and the *y* values are its *range*. For example, the domain of the relation R={(2, 4), (–5, 3), (1, –2)} is {2, –5, 1}, and the range is {4, 3, –2}.

You can graph a linear equation in two ways: *plug and chug* or use *slope-intercept form:* $y = mx + b$. At this point in math, you should definitely know how to use the slope-intercept form, but we give you a quick review of the plug and chug method, because as the equations become more complex, you can use this old standby method to get some key pieces of information.

Graphing using the plug and chug method

Start by picking domain (x) values. Plug them into the equation to solve for the range (y) values. For linear equations, after you plot these points (x, y) on the coordinate plane, you can connect the dots to make a line. The process also works if you choose range values first, then plug in to find the corresponding domain values. This is a helpful method to find *intercepts,* the points that fall on the x or y axes. To find the x-intercept (x, 0), plug in 0 for y and solve for x. To find the y-intercept (0, y), plug in 0 for x and solve for y. For example, to find the intercepts of the linear equation $2x + 3y = 12$, start by plugging in 0 for y: $2x + 3(0) = 12$. Then, using properties of numbers, solve for x: $2x + 0 = 12$, $2x = 12$, $x = 6$. So the x-intercept is (6, 0). For the y-intercept, plug in 0 for x and solve for y: $2(0) + 3y = 12$, $0 + 3y = 12$, $3y = 12$, $y = 4$. Therefore, the y-intercept is (0, 4). At this point, you can plot those two points and connect them to graph the line ($2x + 3y = 12$), because, as you learned in geometry, two points make a line. See the resulting graph in Figure 1-1.

Figure 1-1:
Graph of
$2x + 3y = 12$.

Graphing using the slope-intercept form

The slope-intercept form of a linear equation gives a great deal of helpful information in a cute little package. The equation $y = mx + b$ immediately gives you the y-intercept (b) that you worked to find in the plug and chug method; it also gives you the slope (m). Slope is a fraction that gives you the rise over the run. To change equations that aren't written in slope-intercept form, you simply solve for y. For example, if you use the same linear equation as before, $2x + 3y = 12$, you start by subtracting $2x$ from each side: $3y = -2x + 12$. Next, you divide all the terms by 3: $y = -\frac{2}{3}x + 4$. Now that the equation is in slope-intercept form, you know that the y-intercept is 4. You can graph this point on the coordinate plane. Then, you can use the slope to plot the second point. From the slope-intercept equation, you know that the slope is $-\frac{2}{3}$. This tells you that the rise is -2 and the run is 3. From the point (0, 4), plot the point 2 down and 3 to the right. In other words, (3, 2). Lastly, connect the two points to graph the line. Note that this is the exact same graph, just plotted a different way — the resulting graph in Figure 1-2 is identical to Figure 1-1.

Figure 1-2:
Graph of
$y = -\frac{2}{3}x + 4$.

Similar to graphing equalities, graphing inequalities begins with plotting two points by either method. However, because *inequalities* are used for comparisons — greater than, less than, or equal to — you have two more questions to answer after two points are found:

- Is the line *dashed*: < or > or *solid*: ≤ or ≥?
- Do you shade under the line: $y <$ or $y \le$ or above the line: $y >$ or $y \ge$?

Q. Sketch the graph of the inequality: $3x - 2y > 4$.

A. Begin by putting the equation into slope-intercept form. To do this, subtract $3x$ from each side of the equation: $-2y > -3x + 4$. Then divide each term by -2: $y < \frac{3}{2}x - 2$.

Remember that when you multiply or divide an inequality by a negative, you need to reverse the inequality.

From the resulting equation, you can find the y-intercept, -2, and the slope, $(\frac{3}{2})$. Using this information, you can graph two points using the slope-intercept form method. Next, you need to decide the nature of the line (solid or dashed). Because the inequality is not also an equality, the line is dashed. Graph the dashed line, and then you can decide where to shade. Because the inequality is less than, shade below the dashed line, as you see in Figure 1-3.

Figure 1-3:
Graph of $3x - 2y > 4$.

11. Sketch the graph of $\frac{1}{3}\left(6x + 2y\right) = 16$.

Solve It

12. Sketch the graph of $\frac{5x + 4y}{2} \geq 6$.

Solve It

13. Sketch the graph of $4x + 5y \geq 2\left(3y + 2x + 4\right)$.

Solve It

14. Sketch the graph of $x - 3y = 4 - 2y - y$.

Solve It

Using Graphs to Find Information (Distance, Midpoint, Slope)

Graphs are more than just pretty pictures. From a graph, it's possible to determine two points. From these points, you can determine the distance between them, the midpoint of the segment connecting them, and the slope of the line connecting them. As graphs become more complex in both pre-calculus and calculus, you'll be asked to find and use all three of these pieces of information. Aren't you lucky?

Finding the distance

Distance is how far two things are apart. In this case, you're finding the distance between two points. Knowing how to calculate distance is helpful for when you get to conics (Chapter 12). To find the distance between two points (x_1, y_1) and (x_2, y_2), you can use the following formula:

$$d = \sqrt{\left(x_1 - x_2\right)^2 + \left(y_1 - y_2\right)^2}$$

Calculating the midpoint

The *midpoint*, as you would think, is the middle of a segment. This concept also comes up in conics (Chapter 12) and is ever so useful for all sorts of other pre-calculus calculations. To find the midpoint of those same two points (x_1, y_1) and (x_2, y_2), you just need to average the x and y values and express them as an ordered pair:

$$M = \left(\frac{x_1 + x_2}{2}, \frac{y_1 + y_2}{2} \right)$$

Discovering the slope

Slope is a key concept for linear equations, but it also has applications for trigonometric functions and is essential for differential calculus. *Slope* describes the steepness of a line on the coordinate plane (think of a ski slope). To find the slope of two points (x_1, y_1) and (x_2, y_2), you can use the following formula:

$$m = \frac{y_2 - y_1}{x_2 - x_1}$$

Positive slopes move up and to the right $\left(\diagup_+ \right)$ or down and to the left $\left(\diagup_- \right)$. Negative slopes move down and to the right $\left(\diagdown_+ \right)$ or up and to the left $\left(\diagdown_- \right)$. Horizontal lines have a slope of 0, and vertical lines have an undefined slope.

Q. Find the distance, slope, and midpoint of $\overline{AB}$ in Figure 1-4.

Figure 1-4:
Segment
AB.

A. **The distance is $\sqrt{65}$, the slope is ¹⁄₇, and the midpoint is M = (³⁄₂, 1).** First, plug the x and y values into the distance formula. Then, following the order of operations, simplify the terms under the radical. (Keep in mind those implied parentheses of the radical itself.) It should look something like this:

$$d = \sqrt{\left(5-\left(-2\right)\right)^2 + \left(3-\left(-1\right)\right)^2} = \sqrt{\left(5+2\right)^2 + \left(3+1\right)^2} = \sqrt{\left(7\right)^2 + \left(4\right)^2} = \sqrt{49+16} = \sqrt{65}$$

Because 65 doesn't contain any perfect squares as factors, this is as simple as you can get.

To find the midpoint, plug the points into the midpoint equation. Again, simplify using order of operations.

$$M = \left(\frac{5+\left(-2\right)}{2}, \frac{3+\left(-1\right)}{2}\right) = \left(\frac{3}{2}, \frac{2}{2}\right) = \left(\frac{3}{2}, 1\right)$$

To find the slope, use the formula and plug in your x and y values. Using order of operations, simplify:

$$m = \frac{-1-3}{-2-5} = \frac{-4}{-7} = \frac{4}{7}$$

---

15. Find the distance of segment CD, where C is (–2, 4) and D is (3, –1).

Solve It

16. Find the midpoint of segment EF, where E is (3, –5) and F is (7, 5).

Solve It

17. Find the slope of line GH, where G is
(−3, −5) and H is (−3, 4).

Solve It

18. Find the perimeter of triangle CAT.

Solve It

19. Find the center of the rectangle NEAT.

Solve It

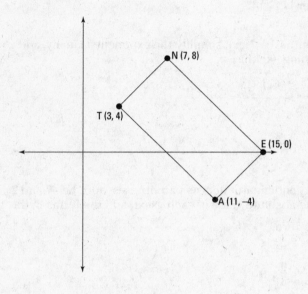

20. Determine whether triangle DOG is a right triangle.

Solve It

Answers to Problems on Fundamentals

1   Simplify $\dfrac{3\sqrt{\left(4-6\right)^2+\left(2-(-1)\right)^2}}{\left|-3-(-1)\right|}$. **The answer is $\dfrac{3\sqrt{13}}{2}$.**

Start by simplifying everything in the parentheses. Next, simplify the exponents. Finally, add the remaining terms. It should look something like this:

$$\frac{3\sqrt{\left(4-6\right)^2+\left(2-(-1)\right)^2}}{\left|-3-(-1)\right|}=\frac{3\sqrt{\left(-2\right)^2+\left(2+1\right)^2}}{\left|-3+1\right|}=\frac{3\sqrt{4+\left(3\right)^2}}{\left|-2\right|}=\frac{3\sqrt{4+9}}{2}=\frac{3\sqrt{13}}{2}$$

2   Simplify $\dfrac{\left|-3\right|-\left|2\right|+(-1)}{\left|-7+2\right|}$. **The answer is 0.**

Recognizing that the absolute value in the denominator acts as parentheses, add the −7 and 2 inside there first. Then, you can rewrite the absolute value of each. Next, add the terms in the numerator. Finally, recognize that $\dfrac{0}{5}$ equals zero.

$$\frac{\left|-3\right|-\left|2\right|+(-1)}{\left|-7+2\right|}=\frac{\left|-3\right|-\left|2\right|+(-1)}{\left|-5\right|}=\frac{3-2+(-1)}{5}=\frac{0}{5}=0$$

3   Simplify $(2^3-3^2)^4(-5)$. **The answer is −5.**

Begin by simplifying the exponents in the parentheses. Next, simplify the parentheses by subtracting 9 from 8. Then, simplify the −1 to the 4th power. Finally, multiply the resulting 1 by −5.

$$(2^3-3^2)^4(-5)=(8-9)^4(-5)=(-1)^4(-5)=1(-5)=-5$$

4   Simplify $\dfrac{\left|51-4+6\right|}{3\left(-\dfrac{1}{6}+\dfrac{1}{3}\right)-\dfrac{1}{2}}$. **The answer is undefined.**

Start by simplifying the parentheses. To do this, subtract 4 from 1 in the numerator and find a common denominator for the fractions in the denominator in order to add them. Next, multiply the terms in the numerator and denominator. Then, add the terms in the absolute value bars in the numerator and subtract the terms in the denominator. Take the absolute value of −9 to simplify the numerator. Finally, remember that you can't have 0 in the denominator; therefore, the resulting fraction ⁹⁄₀ is undefined.

$$\frac{\left|51-4+6\right|}{3\left(-\dfrac{1}{6}+\dfrac{1}{3}\right)-\dfrac{1}{2}}=\frac{\left|5(-3)+6\right|}{3\left(\dfrac{1}{6}\right)-\dfrac{1}{2}}=\frac{\left|-15+6\right|}{\dfrac{1}{2}-\dfrac{1}{2}}=\frac{\left|-9\right|}{0}=\text{Undefined}$$

5   Solve $3-6[2-4x(x+3)]=3x(8x+12)+27$. **The answer is $x=1$.**

Lots of parentheses in this one! Get rid of them by distributing terms. Start by distributing the −4x on the left side over (x + 3) and, on the right side, 3x over (8x + 12). This gives you $3-6[2-4x^2-12x]=24x^2+36x+27$. Then distribute the −6 over the remaining parentheses on the left side of the equation: $3-12+24x^2+72x=24x^2+36x+27$. Combine like terms on the left side: $-9+24x^2+72x=24x^2+36x+27$. To isolate x onto one side, subtract $24x^2$ from each side to get $-9+72x=36x+27$. Subtracting 36x from each side gives you $-9+36x=27$. Adding 9 to both sides results in $36x=36$. Finally, dividing both sides by 36 leaves you with your solution: $x=1$.

6 Solve $\dfrac{x}{2} + \dfrac{x-2}{4} = \dfrac{x+4}{2}$. **The answer is $x = 10$.**

Don't let those fractions intimidate you! Start by multiplying through by the common denominator, 4. This eliminates the fractions altogether. Now, just solve like normal, combining like terms, and isolating x. It should look something like this:

$\dfrac{x}{2} + \dfrac{x-2}{4} = \dfrac{x+4}{2}$; $4\left[\dfrac{x}{2} + \dfrac{x-2}{4} = \dfrac{x+4}{2}\right]$; $2x + x - 2 = 2x + 8$; $3x - 2 = 2x + 8$; $3x = 2x + 10$; $x = 10$

7 Solve $|x - 3| + |3x + 2| = 4$. **The answer is $x = -\tfrac{3}{4}, -\tfrac{1}{2}$.**

Okay, this one is really tricky! Two absolute value terms, oh my! Relax. Just remember that absolute value means distance from 0, so you have to consider all the possibilities to solve this problem. In other words, you have to consider and try four different possibilities: both absolute values are positive, both are negative, the first is positive and the second is negative, and the first is negative and the second is positive.

Not all these possibilities are going to work. As you calculate these possibilities, you may create what math people call *extraneous solutions*. These aren't solutions at all — they're false solutions that don't work in the original equation. You create extraneous solutions when you change the format of an equation, as you're going to do here. So to be sure a solution is real and not extraneous, you need to plug your answer into the original equation to check.

Now, try each of the possibilities:

Positive/positive: $(x - 3) + (3x + 2) = 4$, $4x - 1 = 4$, $4x = 5$, $x = \tfrac{5}{4}$. Plugging this back into the original equation, you get $\tfrac{30}{4} = 4$. Nope! You have an extraneous solution.

Negative/negative: $-(x - 3) + -(3x + 2) = 4$, $-x + 3 - 3x - 2 = 4$, $-4x + 1 = 4$, $-4x = 3$, $x = -\tfrac{3}{4}$. Plug it back into the original equation and you get $4 = 4$. *Voilà!* Your first solution.

Positive/negative: $(x - 3) + -(3x + 2) = 4$, $x - 3 - 3x - 2 = 4$, $-2x - 5 = 4$, $-2x = 9$, $x = -\tfrac{9}{2}$. Put it back into the original equation and you get $12 = 4$. Nope, again — another extraneous solution.

Negative/positive: $-(x - 3) + (3x + 2) = 4$, $-x + 3 + 3x + 2 = 4$, $2x + 5 = 4$, $2x = -1$, $x = -\tfrac{1}{2}$. Into the original equation it goes, and you get $4 = 4$. Your second solution.

8 Solve $3 - 4(2 - 3x) = 2(6x + 2)$. **The answer is no solution.**

To solve, distribute over the parentheses on each side: $3 - 8 + 12x = 12x + 4$. Combine like terms: $-5 + 12x = 12x + 4$. Subtract $12x$ from each side and you get $-5 = 4$, which is false. So there is no solution.

9 Solve $2|x - 3| + 12 = 6$. **The answer is no solution.**

Start by isolating the absolute value: $2|x - 3| + 12 = 6$, $2|x - 3| = -6$, $|x - 3| = -3$. Because an absolute value must be positive, there is no solution that would satisfy this equation.

10 Solve $3(2x + 5) + 10 = 2(x + 10) + 4x + 5$. **The answer is all real numbers.**

Begin by distributing over the parentheses on each side: $3(2x + 5) + 10 = 2(x + 10) + 4x + 5$, $6x + 15 + 10 = 2x + 20 + 4x + 5$. Next, combine like terms on each side: $6x + 25 = 6x + 25$. Subtracting $6x$ from each side gives you $25 = 25$. This is a true statement, indicating that all real numbers would satisfy this equation.

11 Sketch the graph of $\tfrac{1}{3}(6x + 2y) = 16$. **See the graph for the answer.**

Using slope-intercept form, you start by multiplying both sides of the equation by the inverse of $\tfrac{2}{3}$, which is $\tfrac{3}{2}$: $\tfrac{3}{2} \cdot \tfrac{2}{3}(6x + 2y) = \tfrac{3}{2} \cdot 16$. This leaves you with $6x + 2y = 12$. Next, solve for y by subtracting $6x$ from each side and dividing by 2: $2y = -6x + 12$, $y = -3x + 6$. Now, because it's in slope-intercept form, you can identify the slope (-3) and y intercept (6). Use these to graph the equation. Start at the y intercept $(0, 6)$ and move down 3 units and to the right 1 unit. Connect the two points to graph the line.

12 Sketch the graph of $\dfrac{5x + 4y}{2} \geq 6$. **See the graph for the answer.**

Start by multiplying both sides of the equation by 2: $5x + 4y \geq 12$. Next, isolate y by subtracting $5x$ from each side and dividing by 4: $4y \geq -5x + 12$, $y \geq -\frac{5}{4}x + 3$. Now that it's in slope-intercept form, you can graph the inequality. Because it's greater than or equal to, draw a solid line and shade above the line.

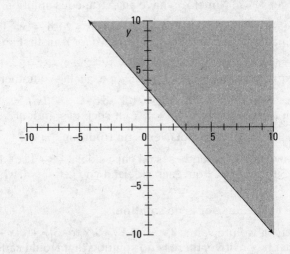

13 Sketch the graph of $4x + 5y \geq 2(3y + 2x + 4)$. **See the graph for the answer.**

Again, start by getting the equation into slope-intercept form. To do this, distribute the 2 on the left side. Next, isolate y by subtracting $4x$ from each side, subtracting y from each side, and then dividing by -1 (don't forget to switch your inequality sign!):

$4x + 5y \geq 2(3y + 2x + 4)$; $4x + 5y \geq 6y + 4x + 8$; $5y \geq 6y + 8$; $-y \geq 8$; $y \leq -8$

Because there's no x term, this indicates that the slope is 0 ($0x$). Therefore, the resulting line is a horizontal line at –8. Because the inequality is less than, you shade below the line.

14 Sketch the graph of $x - 3y = 4 - 2y - y$. **See the graph for the answer.**

Again, simplify to put in slope-intercept form. Combine like terms and add $3y$ to each side.

$x - 3y = 4 - 2y - y$; $x - 3y = 4 - 3y$; $x = 4$

Here, the resulting line is a vertical line at 4.

15 Find the distance of segment CD, where C is (–2, 4) and D is (3, –1). **The answer is $d = 5\sqrt{2}$.**

Using the distance formula, plug in the x and y values: $d = \sqrt{\left(-2-3\right)^2 + \left[4-\left(-1\right)\right]^2}$. Then, simplify using order of operations: $d = \sqrt{\left(-5\right)^2 + \left(5\right)^2}$, $d = \sqrt{25 + 25}$, $d = \sqrt{50}$, $d = 5\sqrt{2}$.

16 Find the midpoint of segment EF, where E is (3, –5) and F is (7, 5). **The answer is *M* = (5, 0).**

Using the midpoint formula, you get $M = \left(\dfrac{3+7}{2}, \dfrac{-5+5}{2} \right)$. Simplify from there: $M = (^{10}\!/_2, ^0\!/_2)$, $M = (5, 0)$.

17 Find the slope of line GH, where G is (–3, –5) and H is (–3, 4). **The answer is *m* = undefined.**

Using the formula for slope, plug in the *x* and *y* values for the two points: $m = \dfrac{-5-4}{-3-(-3)}$. This simplifies to $m = \dfrac{-9}{0}$, which is undefined.

18 Find the perimeter of triangle CAT. **The answer is $8\sqrt{2} + 2\sqrt{26}$.**

To find the perimeter, you need to calculate the distance on each side, which means you have to find CA, AT, and TC. Plugging the values of *x* and *y* for each point into the distance formula, you find that the distances are as follows: CA = $5\sqrt{2}$, AT = $3\sqrt{2}$, and TC = $2\sqrt{26}$. Adding like terms gives you the perimeter of $8\sqrt{2} + 2\sqrt{26}$.

19 Find the center of the rectangle NEAT. **The answer is (9, 2).**

Ah! Think we're being tricky here? Well, the trick is to realize that if you find the midpoint of one of the rectangle's diagonals, you will have identified the center of it. Easy, huh? So, by using

the diagonal NA, you can find the midpoint and thus the center: $M = \left(\dfrac{7+11}{2}, \dfrac{8+(-4)}{2} \right)$. This simplifies to $m = (9, 2)$.

20 Determine whether triangle DOG is a right triangle. **The answer is yes.**

We had to end it with another challenging one. Here you need to remember that right triangles have one set of perpendicular lines (forming that right angle). Also, you need to remember that perpendicular lines have negative reciprocal slopes. In other words, if you multiply their slopes together, you get –1. So, all you have to do to answer this question is find the slopes of the lines that appear to be perpendicular, and if they multiply to equal –1, then you know you have a right triangle. Okay? Then let's go!

Start by finding the slope of DO: $m = \dfrac{4-(-4)}{-3-1}$, $m = -^8\!/_4$, $m = -\dfrac{2}{1}$, or –2. Next, find the slope of

OG: $m = \dfrac{-4-0}{1-9}$, $m = \dfrac{-4}{-8}$, $m = \frac{1}{2}$. Multiplying the two slopes together, you find that, indeed, it does

equal –1, indicating that you have perpendicular lines: $(-2)(\frac{1}{2}) = -1$. Therefore, triangle DOG is a right triangle.

Chapter 2

Get Real!: Wrestling with Real Numbers

In This Chapter

▶ Finding solutions to equations with inequalities

▶ Using interval notation to express inequality

▶ Simplifying radicals and exponents

▶ Rationalizing the denominator

When you build a house, you start by preparing your site and laying your foundation. In Chapter 1, we found and graded the site and started the foundation, but now it's time to make sure that the foundation is set in place before we start building the frame. Pre-calculus, like a sturdy house, has to be based on a solid foundation. In this case, our house is based on Algebra I and II skills. Consider algebra the mortar between your pre-calc bricks. We're going to refresh your memory and cement you with some of those basic skills.

In this chapter, we assume that you know most of your algebra skills well, so we review only the tougher concepts in algebra — the ones that give a lot of our students trouble if they don't review them. In addition to reviewing inequalities, radicals, and exponents, we also introduce a purely pre-calculus idea: interval notation. If you feel confident with the other review sections in this chapter, feel free to skip ahead, but make sure you practice some of the interval notation problems before moving on to Chapter 3. For those of you who aren't sure how solid your cemented foundation is, let's get brickin'!

Solving Inequalities

Solving inequalities is very similar to solving basic equations, which we assume you know solidly by now. There are a few subtle differences, which we'll take the time to review and practice here.

First, remember that an *inequality* is a mathematical sentence indicating that two expressions aren't equal. Inequalities are expressed using the following symbols:

Greater than: >

Greater than or equal to: ≥

Less than: <

Less than or equal to: ≤

Solving equations with inequalities is exactly the same as solving equations with equalities, with one key exception: multiplying and dividing by negative numbers.

When you multiply or divide each side of an inequality by a negative number, you must switch the direction of the inequality symbol. In other words, < becomes > and vice versa.

This is also a good time to put together two key concepts: inequalities and absolute values, or *absolute value inequalities*. With these, you need to remember that absolute values have two possible solutions: one when the quantity in the absolute value bars is positive, and one when it's negative. Therefore, you have to solve for these two possible solutions.

The easiest way to do this is to drop the absolute value bars and apply this simple rule:

$|ax \pm b| < c$ becomes $ax \pm b < c$ AND $ax \pm b > -c$

$|ax \pm b| > c$ becomes $ax \pm b > c$ OR $ax \pm b < -c$

Need an easy way to remember this? Notice the pattern: < is AND, while > is OR. Just think: **"less thAND"** and **"greatOR than."**

The solutions for these absolute value inequalities can be expressed graphically, as follows in Figure 2-1.

Figure 2-1:
Graphical
solution for
$|ax \pm b| > c$.

One more trick those pesky pre-calculus professors may try and pull on you has to do with absolute value inequalities involving negative numbers. You may encounter two possible scenarios:

✔ **If the absolute value is less than or equal to a negative number, a solution doesn't exist.** Because an absolute value must be positive, it can never be less than a negative number. For example, $|2x + 3| \leq -5$ doesn't have a solution.

✔ **If the absolute value is greater than or equal to a negative number, there are infinite solutions, and the answer is all real numbers.** Here, because an absolute value indicates a positive solution and a positive number is always greater than a negative number, an absolute value is always greater than a negative number. For instance, it doesn't matter which number you plug into the equation $|3x - 5| > -2$, you always get a true statement. Therefore, the solution to the statement is all real numbers.

Q. Solve for x in $5 - 2x > 4$.

A. $x < \frac{1}{2}$. Start by subtracting 5 from each side, giving you $-2x > -1$. Next, divide both sides by -2 (don't forget to switch that inequality!), giving you $x < \frac{1}{2}$.

Q. Solve for x in $|4x + 4| - 3 \geq 9$.

A. $x \geq 2$ or $x \leq -4$. First, you have to isolate the absolute value. To do this, add 3 to both sides. Next, drop the absolute value bars and set up your two equations: $4x + 4 \geq 12$ OR $4x + 4 \leq -12$. Solving each algebraically, you get $x \geq 2$ or $x \leq -4$.

1. Solve for x in $|4 - 2x| > 12$.

Solve It

2. Solve for x in $x^2 - 5x - 20 > 4$.

Solve It

3. Solve for x in $|2x + 16| + 15 > 5$.

Solve It

4. Solve for x in $x^3 - 5x > 4x^2$.

Solve It

Expressing Inequality Solutions in Interval Notations

Pre-calc newbie, listen up. This is something you probably didn't experience in algebra. *Interval notation,* although scary sounding, is simply another way of expressing a solution set. Why have another way to write the same thing? Well, this notation is important to know because it's the one most often used in pre-calculus and calculus. And because we know you're incredibly dedicated to becoming a math wizard (uh-huh), you need to know how to cast this spell.

The key to writing a solution in interval notation is to locate the beginning and end of a set of solutions. You can do this by using inequality notation or by visualizing the solution by graphing it. After you locate your key points, you need to decide which type of interval you're dealing with: *open interval* (> or <) or *closed interval* (≥ or ≤).

✔ **Open interval** is indicated by an open circle at a point on a graph and by parentheses in interval notation.

✔ **Closed interval** is indicated by a solid circle at a point on a graph and by brackets for interval notation.

For example, the solution set $-4 \le x < 3$, shown in Figure 2-2, can be rewritten in interval notation as $[-4, 3)$. Another way to think of this solution set is $x \ge -4$ AND $x < 3$.

Figure 2-2:
Graph of
$-4 \le x < 3$.

To indicate a solution set that includes non-overlapping sections (also known as *disjointed* sets), you need to state all the intervals of the solution separated by the word OR. For example, to write the solution set of $x < 2$ or $x \ge 5$ (as shown in Figure 2-3), you need to write both intervals in interval notation: $(-\infty, 2) \cup [5, \infty)$. The symbol between the two sets is the *union* symbol ($\cup$). It means that the solution can belong in either interval.

You always use open interval notation (parentheses) for ∞ or $-\infty$ because they're not real numbers.

Figure 2-3:
Graph of
$x < 2$ or
$x \ge 5$.

Q. Write the solution for $5 - 2x > 4$ in interval notation.

A. $(\frac{1}{2}, \infty)$. We solve this in the previous section, so we just have to write the answer in interval notation.

Q. Graph the interval set $(-2, 3] \cup (5, \infty)$ on a number line.

A. **Figure 2-4.** Start by putting your key points on the number line. Then draw either solid or open circles on your key points, depending on whether they're closed or open intervals. Last, shade.

Figure 2-4:
Graph of
$(-2, 3] \cup$
$(5, \infty)$.

$-2 \qquad\qquad 3 \quad 5$

---

5. Write the solution for the solution of $|2x + 16| + 15 > 5$ in interval notation.

Solve It

6. Write the solution for the solution of $x^3 - 5x > 4x^2$ in interval notation, and graph the solution on a number line.

Solve It

---

7. Graph the interval set $(-\infty, -7) \cup [-5, 2) \cup (4, \infty)$ on a number line.

Solve It

8. Graph the solution of $|2x - 1| \le 3$.

Solve It

Don't Get Drastic with Radicals and Exponents — Just Simplify Them!

Radicals and exponents (also known as *roots* and *powers*) are fundamental algebra concepts. Remember our house? If you don't have a solid foundation of these concepts, you may blow, blow, blow your house down! Here's one area where the foundation needs to be absolutely pre-calc-cemented in place.

To start, we review the definitions of radicals and exponents and their relationship to each other:

A *radical* signifies the root of a number. It's indicated by the radical symbol ($\sqrt{}$). A *root* of a number is a value that must be multiplied by itself to equal that number. For example, the second root (or *square root*) of 9 is 3 because 3 multiplied by itself is 9 ($3 \cdot 3 = 9$). Similarly, the third root (or *cube root*) of 8 is 2 because 2 multiplied by itself two times is 8 ($2 \cdot 2 \cdot 2 = 8$).

An *exponent* is the power of a number. It indicates the number of times a number (the base) is multiplied by itself. For example, 2 to the power of 3 is the same as $2^3 = 2 \cdot 2 \cdot 2 = 8$.

Radicals and exponents are closely related to each other. In fact, they're inverse operations. To solve an equation in which the variable is under a radical, simply take the power of both sides. For example, to solve $\sqrt{x} = 4$, you need to square both sides, giving you $x = 16$. Similarly, you can often solve an equation in which the variable is raised to a power (or has an exponent) by taking the root of both sides. For instance, to solve $x^3 = 27$, you can take the cube root of each side, $\sqrt[3]{x^3} = \sqrt[3]{27}$, to get $x = 3$. You can now use this simple fact to solve equations with radicals and exponents. Who's got the power now, huh?

Sometimes it's easier to solve expressions with radicals and exponents by rewriting them as *rational exponents*, or exponents written as fractions. To do this, remember that the numerator (top number) of the rational exponent is the power, and the denominator (bottom number) is the root.

$$x^{m/n} = \sqrt[n]{x^m} = \left(\sqrt[n]{x}\right)^m$$

For example, you can rewrite $\sqrt[3]{8^2}$ or $\left(\sqrt[3]{8}\right)^2$ as $8^{2/3}$.

Q. Solve for x in $x^2 - 3x^{3/2} - 4x = 0$.

A. **$x = 0, 1, 16$.** Don't let this one scare you! Just remember your basic fraction rules and look for a pattern for factoring. Start by factoring out the GCF (x) from each term.

That leaves you with $x\left(x - 3x^{1/2} - 4\right) = 0$

(remember $x^{3/2} - x^1 = x^{1/2}$, just basic fraction stuff here, nothing fancy). Now, recognize that what's left over in the parentheses is merely a polynomial: $y^2 - 3y - 4 = 0$, then factor this polynomial: $(y - 4)(y + 1) = 0$. When you recognize this, you need only deal with the fractions as the exponents. You can factor the polynomial into

$x\left(x^{1/2} - 4\right)\left(x^{1/2} + 1\right) = 0$. Next, set each factor equal to 0 to find your solutions. It should look something like this:

$$x\left(x^{1/2} - 4\right)\left(x^{1/2} + 1\right) = 0$$

$$x = 0 \qquad x^{1/2} = 4 \qquad x^{1/2} = -1$$
$$\left(x^{1/2}\right)^2 = 4^2 \qquad \left(x^{1/2}\right)^2 = (-1)^2$$
$$x = 16 \qquad x = 1$$

 If you need a refresher of exponential rules, skip ahead to Chapter 5 for a quick review. And for a quick review of solving quadratics, skip ahead to Chapter 4.

Q. Solve for x in $\sqrt{2x - 1} + 4 = x + 2$.

A. **$x = 5$.** Start this one by subtracting 4 from each side, isolating the radical: $\sqrt{2x - 1} = x - 2$. Next, square each side to get rid of the square root: $\left(\sqrt{2x - 1}\right)^2$, which becomes $2x - 1 = (x - 2)(x - 2)$.

 A common mistake is to forget that $(x - 2)^2$ is $(x - 2)(x - 2)$, not $x^2 + 4$!

 Multiply this out using FOIL. A common method for remembering the steps for multiplying two binomials, FOIL stands for F(first), O(outside), I(inside), and L(last). This helps you remember to multiply all the terms.

That leaves you with $2x - 1 = x^2 - 4x + 4$. Bring all terms to one side: $0 = x^2 - 6x + 5$, then factor: $0 = (x - 5)(x - 1)$. Setting each factor equal to 0, you get two possible solutions: $x = 5$ or $x = 1$. Next, plug both solutions back into the original equation to check for extraneous roots (remember those from Chapter 1?). Plugging them in, you find:

$\sqrt{2(5) - 1} + 4 = 5 + 2$	$\sqrt{2(1) - 1} + 4 = 5 + 2$
$\sqrt{10 - 1} + 4 = 7$	$\sqrt{2 - 1} + 4 = 7$
$\sqrt{9} + 4 = 7$	$\sqrt{1} + 4 = 7$
$3 + 4 = 7$	$1 + 4 = 7$
$7 = 7$	$5 \neq 7$

Therefore, $x = 1$ is an extraneous root, and $x = 5$ is the solution!

---

9. Simplify $27^{4/3}$.

Solve It

10. Solve for x in $x^{5/3} - 6x = x^{4/3}$.

Solve It

11. Solve for x in $\sqrt{x-3}-5=0$.

Solve It

12. Solve for x in $x^{8/9}=16x^{2/9}$.

Solve It

13. Solve for x in $\sqrt{x-7}-\sqrt{2x-7}=-2$.

Solve It

14. Solve for x in $x^{2/3}+7x^{1/3}+10=0$.

Solve It

Getting Out of a Sticky Situation, or Rationalizing

Ever find yourself justifying why you deserve a day off after a pre-calc test? It may sound irrational, but rationalizing can come in handy sometimes. To truly simplify a radical expression, you must rationalize your denominators. In other words, you can't leave a radical in the denominator of an expression. In this section, we review and practice *rationalizing the denominator*.

We begin with *monomials* (or one term) in the denominator. For these, it's important to remember that you're dealing with an expression and not an equation, so you need to remember equivalent fractions.

Keep in mind that a monomial is an expression, NOT an equation. You can't simply square the term to find a solution, because you can't counterbalance that action.

Instead, you need to multiply the numerator and denominator by the same term (which is the same as multiplying by 1). For example, if you need to rationalize the expression $\frac{3}{\sqrt{2}}$, you can multiply the expression by $\frac{\sqrt{2}}{\sqrt{2}}$, which equals 1. You then get $\frac{3\sqrt{2}}{2}$.

The same idea works for other radicals, but it requires a little more thinking. For example, if you need to rationalize the expression $\frac{2}{\sqrt[3]{5}}$, you need to multiply the numerator and denominator by $\sqrt[3]{5}$ to the second power, or by $\sqrt[3]{5^2}$, because raising a cube root to the third power cancels the root. After multiplying, you get $\frac{2\sqrt[3]{25}}{5}$.

To rationalize expressions with binomials in the denominator, you must multiply both the numerator and denominator by the conjugate. A *conjugate* is a fancy name for the binomial that, when multiplied by the first binomial, gives you the difference of two squares. It's found by changing the sign of the second term of the binomial. For example, the conjugate of $x + y$ is $x - y$. We know this for sure because when we multiply the two conjugates $(x + y)(x - y)$, we get $x^2 - y^2$, or the difference of two squares.

So to rationalize a denominator with a binomial, start by multiplying the numerator and denominator by the conjugate and simplify. For example, to simplify $\frac{3}{2 - \sqrt{3}}$, multiply the numerator and denominator by $2 + \sqrt{3}$. The steps look like this:

$$\frac{3\left(2 + \sqrt{3}\right)}{\left(2 - \sqrt{3}\right)\left(2 + \sqrt{3}\right)} \quad = \quad \frac{6 + 3\sqrt{3}}{4 - 2\sqrt{3} + 2\sqrt{3} - \sqrt{9}} \quad = \quad \frac{6 + 3\sqrt{3}}{4 - 3} \quad = \quad 6 + 3\sqrt{3}$$

Q. Simplify $\dfrac{12}{\sqrt[3]{9}}$.

A. $x = 4\sqrt[3]{3}$.

To minimize the amount of work you need to do for rationalizing denominators, it's a good idea to rewrite your denominator in factored form to identify the base numbers you're dealing with. So for this problem, we begin by rewriting the denominator: $\dfrac{12}{\sqrt[3]{3^2}}$.

Now, we can multiply the numerator and denominator by $\sqrt[3]{3}$, giving us: $\dfrac{12}{\sqrt[3]{3^2}} \cdot \dfrac{\sqrt[3]{3}}{\sqrt[3]{3}}$.

Simplifying, we get: $\dfrac{12\sqrt[3]{3}}{\sqrt[3]{3^3}} = \dfrac{12\sqrt[3]{3}}{3} = 4\sqrt[3]{3}$.

Q. Simplify $\dfrac{2+\sqrt{5}}{3-\sqrt{5}}$.

A. $x = \dfrac{11+5\sqrt{5}}{4}$. First, multiply the numerator and denominator by the conjugate of the denominator: $\dfrac{\left(2+\sqrt{5}\right)}{\left(3-\sqrt{5}\right)} \cdot \dfrac{\left(3+\sqrt{5}\right)}{\left(3+\sqrt{5}\right)}$.

Be sure to correctly multiply your terms, especially when you're multiplying binomials. A common mistake is to simply distribute the second term to your conjugate in the numerator, but you need to remember that you're multiplying two binomials together: $\dfrac{\left(2+\sqrt{5}\right)}{\left(3-\sqrt{5}\right)} \cdot \dfrac{\left(3+\sqrt{5}\right)}{\left(3+\sqrt{5}\right)}$, not

$\dfrac{2+\sqrt{5}\left(3+\sqrt{5}\right)}{\left(3-\sqrt{5}\right)\left(3+\sqrt{5}\right)}$. Correctly multiplied out using FOIL, you should get:

$\dfrac{6+2\sqrt{5}+3\sqrt{5}+5}{9+3\sqrt{5}-3\sqrt{5}-5}$. Combining like terms, the final answer is $\dfrac{11+5\sqrt{5}}{4}$.

15. Simplify $\sqrt{\dfrac{3}{2x+4}}$.

Solve It

16. Simplify $\dfrac{\sqrt{6}+\sqrt{8}}{\sqrt{10}-\sqrt{2}}$.

Solve It

17. Simplify $\dfrac{3\sqrt[5]{2}}{2\sqrt[5]{18}}$.

Solve It

18. Simplify $\dfrac{8}{4^{2/3}}$.

Solve It

Answers to Problems on Real Numbers

1. Solve for x in $|4 - 2x| > 12$. **The answer is $x > 8$ or $x < -4$.**

Start by dropping the absolute value bars and setting up your two equations: $4 - 2x > 12$ or $4 - 2x < -12$. Solve algebraically (careful when you divide by that negative!): $-2x > 8$ or $-2x < -16$; $x < -4$ or $x > 8$.

2. Solve for x in $x^2 - 5x - 20 > 4$. **The answer is $x > 8$ or $x < -3$.**

First, you need to recognize that we're dealing with a quadratic here. To solve a quadratic, you need to isolate it. (If you need to review quadratics, skip ahead to Chapter 4). Start by subtracting 4 from each side: $x^2 - 5x - 24 > 0$. Next, factor your quadratic: $(x - 8)(x + 3) > 0$. Now what? Don't worry; we'll lead you through this! Setting each factor to 0 gives you your key points: 8 and −3. If you put these on a number line, you can see that you have three possible solutions: less than −3, between −3 and 8, or greater than 8. All you have to do is plug in numbers in each interval to see if you get a positive or negative number. Because you're looking for a solution that's greater than 0, you need a positive result when you multiply your factors, $(x - 8)$ and $(x + 3)$, together. In other words, you want both of your factors to be positive or both of them to be negative. Looking at the number line in the following figure, you see that your solutions are $x > 8$ or $x < -3$. Whew!

$$\begin{array}{ccc} (x-8) \text{ negative} & (x-8) \text{ negative} & (x-8) \text{ positive} \\ (x+3) \text{ negative} & (x+3) \text{ positive} & (x+3) \text{ positive} \end{array}$$

$$\longleftarrow \quad \underset{-3}{|} \qquad \underset{8}{|} \quad \longrightarrow$$

3. Solve for x in $|2x + 16| + 15 > 5$. **The answer is all real numbers.**

Begin by isolating the absolute value by subtracting 15 from each side, giving you $|2x + 16| > -10$. Remember that absolute values are positive and therefore greater than any negative. No matter what you plug in for x, you get a positive number. So, the solution is all real numbers!

4. Solve for x in $x^3 - 5x > 4x^2$. **The answer is $x > 1$ or $0 > x > -5$.**

First, if you need a refresher on solving polynomials and quadratics, skip ahead to Chapter 4. For this problem, start by gathering all your variables to one side of the equation by subtracting $4x^2$ from each side: $x^3 + 4x^2 - 5x > 0$. Next, factor out x from each term: $x(x^2 + 4x - 5) > 0$. Then factor the quadratic: $x(x - 5)(x - 1) > 0$. Setting your factors equal to 0, you can find your key points. Put these points on a number line. Plug in test numbers from each possible section to determine whether the factor would be positive or negative. Then, given that you're looking for a positive solution, think about the possibilities: $(+)(+)(+) = (+), (+)(+)(-) = (-), (-)(+)(-) = (+), (-)(-)(-) = (-)$. Therefore, your solution is $x > 1$ or $0 > x > -5$.

5. Write the solution for the solution of $|2x + 16| + 15 > 5$ in interval notation. **The answer is $(-\infty, \infty)$.**

Recognize this one? We solved it in practice problem 3. The solution is all real numbers, and you write that in interval notation by writing it as infinity to negative infinity. Cool, huh?

6. Write the solution of $x^3 - 5x > 4x^2$ in interval notation, and graph the solution on a number line. **The answer is $(-5, 0) \cup (1, \infty)$.**

Oops, we did it again! This one came from practice problem 4. The graph looks like this:

7 Graph the interval set $(-\infty, -7) \cup [-5, 2) \cup (4, \infty)$ on a number line.

8 Graph the solution of $|2x - 1| \leq 3$.

Start by dropping the absolute value sign and setting up your two equations: $2x - 1 \leq 3$ and $2x - 1 \geq -3$. Then solve each to find your solution: $2x \leq 4$ and $2x \geq -2$; $x \leq 2$ and $x \geq -1$. These can also be rewritten as $-1 \leq x \geq 2$, which can be graphed as follows:

9 Simplify $27^{4/3}$. **The answer is 81.**

First, recognize that you can think of this problem in two ways: $\sqrt[3]{(27)^4}$ and $\left(\sqrt[3]{27}\right)^4$. Either way gives you the correct answer, but one is easier to deal with than the other. Starting with $\sqrt[3]{(27)^4}$, order of operations tells us to take the 27 to the power of 4 first, giving us: $\sqrt[3]{531441}$. Ew! No fun! If we choose to deal with the problem written like this: $\left(\sqrt[3]{27}\right)^4$, then it's much easier. Order of operations here tells us to take the cube root of 27, which is 3, then take 3 to the 4th power, which is 81. Ah . . . much better. By the way, the cube root of 531,441 is 81 as well, but we won't make you memorize it.

10 Solve for x in $x^{5/3} - 6x = x^{4/3}$. **The answer is $x = 0, -8, 27$.**

Begin by bringing all the terms to one side in descending order: $x^{5/3} - x^{4/3} - 6x = 0$. Next, factor out an x from each term: $x\left(x^{2/3} - x^{1/3} - 6\right) = 0$. Then, you can see that the resulting quadratic is similar to $y(y^2 - y - 6)$, which factors into $y(y + 2)(y - 3)$. Similarly, you can factor $x\left(x^{2/3} - x^{1/3} - 6\right) = 0$ into $x\left(x^{1/3} + 2\right)\left(x^{1/3} - 3\right) = 0$. Then, setting each factor equal to 0 and simplifying, you can find your three solutions:

$$x\left(x^{1/3} + 2\right)\left(x^{1/3} - 3\right) = 0$$

$$x = 0 \qquad\qquad x^{1/3} + 2 = 0 \qquad x^{1/3} - 3 = 0$$

$$x^{1/3} = -2 \qquad x^{1/3} = 3$$

$$\left(x^{1/3}\right)^3 = (-2)^3 \qquad \left(x^{1/3}\right)^3 = (3)^3$$

$$x = -8 \qquad\qquad x = 27$$

11 Solve for x in $\sqrt{x - 3} - 5 = 0$. **The answer is $x = 28$.**

Start by isolating the radical by adding 5 to each side, giving you $\sqrt{x - 3} = 5$. Next, square both sides to get rid of the square root. This gives you $\left(\sqrt{x - 3}\right)^2 = 5^2$, which simplifies to $x - 3 = 25$, which is the same as $x = 28$.

12 Solve for x in $x^{8/9} = 16x^{2/9}$. **The answer is $x = 64$.**

First, divide both sides by $x^{2/9}$, giving you: $\dfrac{x^{8/9}}{x^{2/9}} = \dfrac{16x^{2/9}}{x^{2/9}}$. Simplify using your exponent rules

(see Chapter 5 for a refresher!): $x^{8/9 - 2/9} = 16$, $x^{2/3} = 16$. Next, raise each side to the power of $\frac{3}{2}$:

$\left(x^{2/3}\right)^{3/2} = (16)^{3/2}$. Then, simplify the radical $x = \left(\sqrt{16}\right)^{3}$ by taking the square root of 16 and raising

it to the power of 3. You get 64.

13 Solve for x in $\sqrt{x-7} - \sqrt{2x-7} = -2$. **The answer is $x = 8, 16$.**

Begin by isolating one of the radicals: $\sqrt{x-7} = -2 + \sqrt{2x-7}$. Then, square both sides to get rid of

that radical: $\left(\sqrt{x-7}\right)^{2} = \left(-2 + \sqrt{2x-7}\right)^{2}$. Make sure that you multiply your binomials correctly!

$x - 7 = \left(-2 + \sqrt{2x-7}\right)\left(-2 + \sqrt{2x-7}\right)$. Multiplying the terms on the right side of the equation gives

you $x - 7 = 4 - 4\sqrt{2x-7} + 2x - 7$. Next, isolate the remaining radical using basic algebra:

$4\sqrt{2x-7} = x + 4$. Then, you can square both sides again to remove the remaining radical:

$\left(4\sqrt{2x-7}\right)^{2} = (x+4)^{2}$. Using algebra, multiply the two binomials and combine like terms:

$16(2x-7) = (x+4)(x+4)$; $32x - 112 = x^2 + 8x + 16$; $0 = x^2 - 24x + 128$. This quadratic factors into:

$0 = (x-8)(x-16)$. Setting both factors equal to 0, you get two possible solutions: $x = 8$ and

$x = 16$. Plug both back into the original equation and you'll find that both solutions work.

14 Solve for x in $x^{2/3} + 7x^{1/3} + 10 = 0$. **The answer is $x = -8, -125$.**

Start by recognizing that this trinomial is similar to $y^2 + 7y + 10 = 0$, which factors to

$(y + 5)(y + 2) = 0$. Similarly, $x^{2/3} + 7x^{1/3} + 10 = 0$ factors into $\left(x^{1/3} + 5\right)\left(x^{1/3} + 2\right) = 0$. Setting each one

equal to 0, you can easily solve for the solutions by taking each side to the power of 3. In other

words, $x^{1/3} = -5$ becomes $\left(x^{1/3}\right)^{3} = (-5)^{3}$, so $x = -125$, and $x^{1/3} = -2$ becomes $\left(x^{1/3}\right)^{3} = (-2)^{3}$, so $x = -8$.

15 Simplify $\sqrt{\dfrac{3}{2x+4}}$. **The answer is $\dfrac{\sqrt{6x+12}}{2x+4}$.**

First, you need to separate the fraction into two radicals: one in the numerator and one in the

denominator: $\dfrac{\sqrt{3}}{\sqrt{2x+4}}$. Now, multiply the numerator and denominator by the square root in the

denominator: $\dfrac{\sqrt{3}}{\sqrt{2x+4}} \cdot \dfrac{\sqrt{2x+4}}{\sqrt{2x+4}}$. Did we trick you here? This one doesn't require the use of a

conjugate because there isn't another term added to the radical. Simplify the numerator by

multiplying the radicals: $\dfrac{\sqrt{3(2x+4)}}{2x+4} = \dfrac{\sqrt{6x+12}}{2x+4}$.

16 Simplify $\dfrac{\sqrt{6}+\sqrt{8}}{\sqrt{10}-\sqrt{2}}$. **The answer is** $\dfrac{\sqrt{15}+\sqrt{3}+2\sqrt{5}+2}{4}$.

Start by multiplying the numerator and denominator by the conjugate of the denominator: $\dfrac{\left(\sqrt{6}+\sqrt{8}\right)}{\left(\sqrt{10}-\sqrt{2}\right)}\cdot\dfrac{\left(\sqrt{10}+\sqrt{2}\right)}{\left(\sqrt{10}+\sqrt{2}\right)}$. Next, multiply the binomials in the numerator and denominator using FOIL: $\dfrac{\sqrt{60}+\sqrt{12}+\sqrt{80}+\sqrt{16}}{10-2}$. Then, simplify each radical: $\dfrac{2\sqrt{15}+2\sqrt{3}+4\sqrt{5}+4}{8}$. Finally, because each term in the numerator and denominator is divisible by 2, divide both by 2: $\dfrac{\sqrt{15}+\sqrt{3}+2\sqrt{5}+2}{4}$.

17 Simplify $\dfrac{3\sqrt[5]{2}}{2\sqrt[5]{18}}$. **The answer is** $\dfrac{\sqrt[5]{3^3}}{2}$.

Begin by factoring the denominator: $\dfrac{3\sqrt[5]{2}}{2\sqrt[5]{2\cdot3^2}}=\dfrac{3\sqrt[5]{2}}{2\sqrt[5]{2}\cdot\sqrt[5]{3^2}}$. Notice the $2\sqrt[5]{2}$ in both the numerator and denominator? Cancel them! Yea — one less term to worry about! Next, multiply the numerator and denominator of $\dfrac{3}{2\sqrt[5]{3^2}}$ by $\sqrt[5]{3^3}$ to eliminate the radical in the denominator: $\dfrac{3}{2\sqrt[5]{3^2}}\cdot\dfrac{\sqrt[5]{3^3}}{\sqrt[5]{3^3}}$. Multiply through: $\dfrac{3\sqrt[5]{3^3}}{2\sqrt[5]{3^5}}=\dfrac{3\sqrt[5]{3^3}}{2\cdot3}$ and then cancel the 3 from the numerator and denominator to get your final answer: $\dfrac{\sqrt[5]{3^3}}{2}$.

18 Simplify $\dfrac{8}{4^{2/3}}$. **The answer is** $2\sqrt[3]{4}$.

Start by changing the fractional exponent into a radical: $\dfrac{8}{\sqrt[3]{4^2}}$. Then, multiply the numerator and denominator by one more cube root of 4: $\dfrac{8}{\sqrt[3]{4^2}}\cdot\dfrac{\sqrt[3]{4}}{\sqrt[3]{4}}$. Multiply: $\dfrac{8\sqrt[3]{4}}{4}$, and simplify the fraction: $2\sqrt[3]{4}$.

Chapter 3

Understanding the Function of Functions

You've seen the *coordinate plane,* where two number lines meet at a 90° angle. You know that the horizontal axis is called the *x*-axis, and the vertical one is called the *y*-axis. You also know that each point, or *ordered pair,* on the plane is named (x, y). But did you know that a *relation* is a set of ordered pairs? The *domain* of the relation is the set of all the *x* values, and the *range* is the set of all the *y* values. ***Note:*** If you ever run into different variables (like *m* and *n*), domain and range will be based alphabetically.

A *function* is a relation where every *x* in the domain pairs with one (and only one) *y* in the range. The symbol for a function is $f(x)$, read "function of *x*," or simply "*f* of *x*." Think of a function as a computer. Domain is input and range is output. You can't put input in a computer and get out different outputs; otherwise, your computer would be broken. If you had the world's simplest computer and all it did was multiply by 3, if you input 2 in the computer, you better get an output of 6. Domain and range have a similarly correspondent relationship. We explore the idea of functions and some properties of them in this chapter. Now get computing!

Battling Out Even versus Odd

If you've ever taken an art class, you've probably heard the term *symmetry*. It means that the picture is balanced, with equal or similar parts on both sides of the painting. A graph can be symmetrical as well. Algebra has three different types of symmetry:

✔ *Y*-axis symmetry: Each point on the left side of the *y*-axis is mirrored by a point on the right side, and vice versa.

✔ *X*-axis symmetry: Each point above the *x*-axis is mirrored by a point below it, and vice versa.

✔ Origin symmetry: If you turn the graph upside down, it looks exactly the same.

In pre-calculus, functions take this idea of symmetry and use different terms to describe the same idea. A function whose graph is symmetrical with respect to the *y*-axis is called an *even function*. Basically, each input *x* and the opposite input –*x* give the same *y* value. In symbols, textbooks write that $f(x) = f(-x)$. A function whose graph is symmetrical with respect to the origin is called an *odd function*. In plain English, each *x* value gives a *y* value, and its opposite –*x* gives the opposite –*y*. This means that $f(x) = -f(x)$.

EXAMPLE

Q. Determine whether f(x) = $x^4 - x^2$ is even, odd, or neither.

A. **This function is even.** Replace x with $-x$ in the equation and see what happens: f($-x$) = $(-x)^4 - (-x)^2$. A negative number to an even power is a positive number. So f($-x$) = $x^4 - x^2$. Because you get the same exact function as the original one, this function is even.

---

1. Is f(x) = $x^3 - 1$ even, odd, or neither?

Solve It

2. Determine whether the given graph is even, odd, or neither.

Solve It

---

3. Sketch half the graph of f(x) = $\sqrt{(x^2 - 4)}$ and use symmetry to complete the graph.

Solve It

4. Sketch half the graph of f(x) = $4x^3$ and use symmetry to complete the graph.

Solve It

Leaving the Nest: Transforming Parent Graphs

You see certain functions over and over again in pre-calc, and sometimes you'll be asked to graph them. The plug and chug method for graphing works for any equation (including functions), but the more complicated the function, the longer it takes you to graph it using this method. The basic graph, in any case, is called the *parent graph*.

Common parent graphs include quadratic functions, square roots, absolute values, cubics, and cube roots. Moving these basic graphs around the coordinate plane is known as *transforming the function* and is easier than the plug and chug method. Several types of transformations of functions exist:

- ✔ Horizontal transformations
- ✔ Vertical transformations
- ✔ Reflections
- ✔ Horizontal translations
- ✔ Vertical translations

In this section, we take a look at each parent function and then show you how to transform them. ***Note:*** Even though in most sections we take a look at only one function when discussing the transformations, the rules apply to *all functions* in the same way. So if we talk about a quadratic function in the section on vertical transformations, that's not the only function that has vertical transformations — they all do.

Quadratic functions

Quadratic functions are second degree equations. The highest exponent on any one variable is two. The parent quadratic function is $f(x) = x^2$. Its graph is known as a *parabola;* we talk in depth about parabolas in Chapter 12. Begin the graph at the *vertex* $(0, 0)$, and to get to the next point, move over 1, up 1; the next point from there is over 1, up 3; the next point is over 1, up 5. You always move over 1 and up the next odd number. See Figure 3-1 for the graph of the parent quadratic function. The parent quadratic function is an even graph — it's symmetrical with respect to the *y*-axis.

Figure 3-1:
Graphing the parent quadratic function.

Square root functions

The parent *square root function* is $g(x) = \sqrt{x}$. This graph looks like half a parabola, turned on its side. You can't have x values that are negative because you can't square root a negative number, and $f(x)$ is also positive because $\sqrt{x}$ represents the *principle*, or positive root. The graph begins at the origin $(0, 0)$ and then moves up 1, over 1; up 1, over 3; up 1, over 5; and so on. The graph of the parent square root function can be seen in Figure 3-2.

Figure 3-2:
The parent
square root
function.

Absolute value functions

The absolute value parent graph is determined by $h(x) = |x|$. You should recognize the absolute value bars and know that this figure represents distance, so it always gives a positive output. This parent function also starts at $(0, 0)$ and then always moves over 1, up 1. Its graph can be seen in Figure 3-3. The parent absolute value function is also an even function.

Figure 3-3:
The
absolute
value parent
function.

Cubic functions

A *cubic function* is one where the highest degree on a variable is three: $p(x) = x^3$ is the parent function in this case. The parent cubic function is an odd function; if you turn it upside down it looks exactly the same. If you start at the origin $(0, 0)$ and mark a point, the point immediately to the left is over 1 and down 1, and the point immediately to the right is over 1 and up 1. This graph is shown in Figure 3-4.

Figure 3-4:
The cubic function's parent graph.

Cube root functions

Cube root functions are related to cubic functions in a similar way that quadratic and square root functions are. The parent graph is $r(x) = \sqrt[3]{x}$. The graph also starts at the origin and moves right 1, up 1 and left 1, down 1, but the graph is longer than it is tall. Take a look at Figure 3-5 for this parent graph.

Figure 3-5:
The cube root parent function.

Vertical transformations

In any function, the method of moving up to get to the next point is known as the *vertical transformation*. Some teachers may explain this concept as *amplitude*, but that's technically incorrect because the functions we've discussed so far keep going up forever, so they don't technically have an amplitude or height. Multiplying any function by a constant changes a graph's vertical transformation. This is written as $a \cdot f(x)$. Think of this as a vertical stretch or shrink. A coefficient between 0 and 1 is a *shrink*, and a coefficient greater than 1 is a *stretch*.

For example, $f(x) = 2x^2$ multiplies each up value by 2. From the vertex, you used to move over 1, up 1, but now you move over 1, up 2. After that, you used to move over 1, up 3, but now you move over 1, up 6. This keeps going for each point, which is why some teachers call it amplitude — the height of each individual point is affected.

For another example, if $g(x) = \frac{1}{4}x^2$, you move in this manner: From the vertex, go over 1, up $\frac{1}{4}$; over 1, up $\frac{3}{4}$; over 1, up $\frac{5}{4}$; and so on. The graphs of $f(x)$ and $g(x)$ are shown in Figure 3-6.

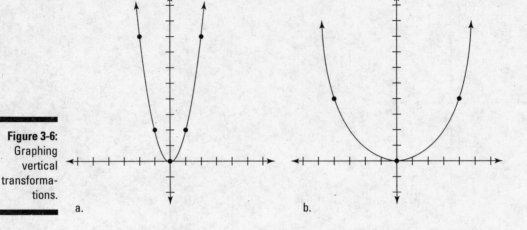

Figure 3-6:
Graphing
vertical
transforma-
tions.

a. b.

Note: A negative coefficient is actually a reflection, so we discuss that type of transformation in the upcoming "Reflections" section.

Horizontal transformations

If the vertical transformation affects the up values in the function, then a *horizontal transformation* affects the over values. Horizontal transformations stretch or shrink the graph along the *x*-axis. This time, the coefficient is inside the function: f[*c*(*x*)].

$h(x) = 3|x|$ is a vertical transformation; $h(x) = |3x|$ is a horizontal transformation.

$f(x) = 4x^2$ is a vertical transformation; $f(x) = (4x)^2$ is a horizontal transformation.

Set the expression inside the function equal to the parent function's normal horizontal transformation and solve for *x* to find the new value. If $g(x) = |3x|$, then set $3x = 1$ and solve for $x = \frac{1}{3}$. From the origin, you move over $\frac{1}{3}$, up 1; over $\frac{1}{3}$, up 1; over $\frac{1}{3}$, up 1.

Translations

Moving a graph horizontally or vertically on the coordinate plane is called a *translation*. Every point on the parent graph is moved right, left, up, or down. Here we take a closer look at each kind of translation, or *shift*.

Horizontal shifts

Adding or subtracting a number inside the function s grouping device is a *horizontal shift*. A horizontal shift is always written in the form f(*x* – *h*) such that the horizontal shift is the opposite of what it appears to be.

$h(x) = \sqrt{(x-2)}$ moves the parent square root function to the right by 2.

$h(x) = \sqrt{(x+3)}$ moves the parent square root function to the left by 3.

Vertical shifts

Adding or subtracting a constant outside of the function's grouping device is a *vertical shift*. These types of transformations are written as f(x) + v, where v is the vertical shift.

p(x) = x^3 – 1 moves the parent cubic function down by 1.

p(x) = x^3 + 4 moves the parent cubic function up by 4.

Reflections

Reflections take the parent function and reflect it over a horizontal or vertical line. When the vertical transformation coefficient is negative, the function is flipped upside down over a horizontal line. For example, f(x) = $-5x^2$ affects the vertical transformation by a factor of 5 and turns the graph upside down at the same time. If the horizontal transformation is negative, the function is flipped backwards over a vertical line:

h(x) = $\sqrt{(-x)}$ turns the function to the left instead of to the right.

Combinations of transformations

Putting some or all the transformations into one function is itself a transformation. Putting all of them together into one expression looks like this: $a \cdot f[c(x - h)] + v$ where

a is the vertical transformation

c is the horizontal transformation

h is the horizontal translation

v is the vertical translation

We recommend doing the translations first and then doing the transformations.

Q. Graph the function f(x) = $(x - 3)^2$ by transforming the parent graph.

A. **See the graph.** This transformation is done in one step. Because the constant is subtracting inside the quadratic function, you recognize it as a horizontal shift to the right by 3. Take the parent quadratic function and move each point to the right by 3, as shown in the graph.

f(x) = $(x - 3)^2$

_Q._ Sketch the graph of $g(x) = \sqrt{3-x} + 1$ by transforming the parent function.

_A._ **See the graph.** This one takes some work before you can begin graphing it. It must be rewritten in the proper form to recognize the various transformations to the parent square root function. First, rewrite the stuff inside the square root so that it's in the right order: $g(x) = \sqrt{-x+3} + 1$. Next, factor out the leading coefficient to get the horizontal transformation: $g(x) = \sqrt{-1(x+3)} + 1$. This means that the graph is flipped horizontally. Notice that factoring out the coefficient affects the horizontal translation — it's to the right by 3, and the vertical translation is up 1. These transformations can be seen in the final graph.

$g(x) = \sqrt{3-x} + 1$

---

5. Graph the function $a(x) = -2(x-1)^2$.

Solve It

6. Graph the function $b(x) = |x+4| - 1$.

Solve It

7. Graph the function $c(x) = \sqrt{x+3}$.

Solve It

8. Graph the function $f(x) = -x^2 - 6x$.

Solve It

Given the graph of the function g in Figure 3-7, sketch the graph of the functions in Problems 9 and 10.

Figure 3-7:
The given
parent func-
tion g(x).

9. $g(x) - 3$.

Solve It

10. $2g(x - 1)$.

Solve It

Lucid Thinking? Graphing Rational Functions

A *rational function* is one where the variable is in a fraction's denominator. You know by now that when the denominator of a fraction is 0, the result is undefined. The same is true for rational functions. Because the denominator has a variable, it may be possible that certain values of x will make the denominator 0. If the function has values that make it *undefined,* the graph will have a vertical asymptote.

To find the vertical asymptote, if there is one, set the denominator equal to 0 and solve. This is a vertical line that the graph will never cross. Some rational functions also have a *horizontal asymptote* as well. A graph *shouldn't* ever cross a horizontal asymptote, but in some cases it will, so don't freak out if that happens. (The second upcoming example is a rational function where the graph crosses its horizontal asymptote.)

To find the horizontal asymptote, take a look at both the numerator's degree and the denominator's degree. (If you've forgotten how to find the degree of a polynomial, see Chapter 4.) Here are the three possibilities for horizontal asymptotes:

- ✔ **The degree of the denominator is greater:** This means that the bottom of the fraction is getting bigger, faster, and the fraction will go to 0 as x gets larger. Your horizontal asymptote is the x-axis, or $y = 0$.

- ✔ **The degree of both is the same:** This means that the top and bottom of the fraction are moving at the same rate. The quotient of the leading coefficients gives you the horizontal asymptote.

- ✔ **The degree of the numerator is greater:** This means that the top of the fraction is getting bigger, faster. A really big number divided by a smaller number is still a pretty big number. In short, as x gets larger, so will y, and there's no horizontal asymptote. Instead, there's an *oblique asymptote*. To find it, find the quotient by dividing the denominator into the numerator. The function that you find is the one that you graph as an oblique asymptote.

EXAMPLE

Q. Graph the function $f(x) = \dfrac{3x-1}{4-x}$.

A. **See the graph.** First, find the vertical asymptote (if there is one) by setting the denominator equal to 0 and solving. If $4 - x = 0$, then $x = 4$. Draw a coordinate plane and add in a dotted vertical line at $x = 4$ to mark your vertical asymptote. Now, look at the numerator and the denominator; the degree on each is one. Divide the leading coefficients to find the horizontal asymptote. In this case, the numerator's leading coefficient is 3 and the denominator's is –1. This means your horizontal asymptote is $y = \frac{3}{-1} = -3$. Now that you have both asymptotes, use them to help you get the graph. The vertical asymptote divides your domain into two intervals: $(-\infty, 4)$ and $(4, \infty)$. Pick a couple of x-values on each interval and plug them into the function to determine whether the graph lives above or below the horizontal asymptote. For example, if $x = -5$, then $y = -1.77$; and if $x = 0$, then $y = -0.25$. If you graph those two points, you see that they're both above the horizontal asymptote. Keep checking points until you have a good idea of what the graph looks like.

Q. Graph the function $g(x) = \dfrac{2x-6}{x^2+4}$.

A. **See the graph.** This time, when you try to find the vertical asymptote, you notice that $x^2 + 4 = 0$ doesn't have a solution because $x^2 = -4$ has no solution (in the real numbers anyway). Also notice that because the denominator has a bigger degree, the horizontal asymptote is the x-axis, or $y = 0$. However, also notice that by setting the numerator equal to 0, you do get a solution: $2x - 6 = 0$; $2x = 6$; $x = 3$. This means the graph crosses the x-axis at $x = 3$ even though it's not supposed to. Because there's no vertical asymptote, use this value to give you the intervals to look at to get the graph. On the first interval $(-\infty, 3)$, y is negative and the whole graph is below the horizontal asymptote. On the next interval $(3, \infty)$, y happens to be positive and the function is above the horizontal asymptote. If you pick x values bigger than 3 that keep getting bigger, you see y increase slowly and then decrease again and get closer and closer to 0. This gives you the graph of this function.

11. Graph the function $q(x) = \dfrac{-2}{(x-4)(x+5)}$.

Solve It

12. Graph the function $r(x) = \dfrac{x+3}{x^2-x-6}$.

Solve It

13. Graph the function $t(x) = \dfrac{x^2-5x}{x^2-4x-21}$.

Solve It

14. Graph the function $u(x) = \dfrac{x^2-10x-24}{x+1}$.

Solve It

Picking Up the Pieces: Graphing Piece-Wise Functions

A *piece-wise* function is called that because it's broken into pieces. A piece-wise function actually contains several functions, each defined on a restricted interval. The output depends on what the input is. The graphs of these functions may look like they've literally been broken into pieces. Because of this broken quality, a piece-wise function that jumps is called *discontinuous*.

Q. Graph f(x) = $\begin{cases} x^2 + 2 \text{ if } x \le 1 \\ 3x - 1 \text{ if } x > 1 \end{cases}$

A. **See the graph.** This function has been broken into two pieces: When $x \le 1$, the function follows the graph of the quadratic function, and when $x > 1$, the function follows the graph of the linear function. Notice the hole in this second piece of the graph to indicate that the point isn't actually there. The graph is shown in Figure 3-8.

Figure 3-8:
The graph of one piece-wise function.

---

15. Graph g(x) = $\begin{cases} \sqrt{x + 3} \text{ if } x \le -1 \\ (x + 3)^2 \text{ if } x > -1 \end{cases}$

Solve It

16. Graph h(x) = $\begin{cases} \frac{1}{2}x - 4 \text{ if } x \le -2 \\ 3x + 3 \text{ if } -2 < x < 2 \\ 4 - x \text{ if } x \ge 2 \end{cases}$

Solve It

17. Graph $m(x) = \begin{cases} x^3 + 2 \text{ if } x < 0 \\ x^2 + 2 \text{ if } 0 \le x < 2 \\ x + 2 \text{ if } x \ge 2 \end{cases}$

Solve It

18. Graph $n(x) = \begin{cases} |x - 1| \text{ if } x < -3 \\ -3 \text{ if } x = 3 \\ |x| - 1 \text{ if } x > -3 \end{cases}$

Solve It

Operating on Functions: No Scalpel Necessary

You've come to know (and maybe to love, too?) four basic operations in math: addition, subtraction, multiplication, and division. Well, in pre-calc, you take functions and add, subtract, multiply, and divide them. By doing this, you create a brand new function. This is sometimes called *combining functions*. In general, you probably won't be asked to graph a combined function because it usually won't be based on any of the parent functions. If you were asked to graph a combined function, you'd have to plug and chug your way through it by picking plenty of *x*-values to make sure you get an accurate representation of the graph. You may also be asked to find one specific value for a combined function — you get an *x* value and you just plug it in and see what happens. Exciting, we know!

For all questions in this section, you use three functions:

$$f(x) = x^2 - 6x + 2$$
$$g(x) = 2x^2 - 5x$$
$$h(x) = \sqrt{3x + 2}$$

Q. Find $(f - g)(x)$.

A. $(f - g)(x) = -x^2 - x + 2$. Because these two functions are both polynomials, solving this is really about collecting like terms and subtracting them. Just be sure to watch your negative signs! $(f - g)(x) = (x^2 - 6x + 2) - (2x^2 - 5x) = x^2 - 6x + 2 - 2x^2 + 5x = -x^2 - x + 2$. No problem!

19. Find $(f + h)(x)$.

Solve It

20. Find $(fg)(x)$.

Solve It

21. Find $(h/g)(x)$. Does this new function have any undefined values?

Solve It

22. Find $(g + h)(2)$.

Solve It

Evaluating Composition of Functions

Placing one function inside of another (or, actually, in itself) is called a *composition of functions*. If you have two functions f(x) and g(x), then the composition f(g(x)) takes g and places it inside of f. This is also written as $(f \circ g)(x)$, and it's basically read right to left; the g function goes into the f function.

You use the same three functions from the last section here as well:

$$f(x) = x^2 - 6x + 2$$
$$g(x) = 2x^2 - 5x$$
$$h(x) = \sqrt{3x + 2}$$

_Q._ Find f(h(x)).

_A._ $3x + 4 - 6\sqrt{3x + 2}$. Start by substituting the entire h function for every x in the f function: $\sqrt{3x + 2}^2 - 6\sqrt{3x + 2} + 2$. A square root and a square cancel each other: $3x + 2 - 6\sqrt{3x + 2} + 2$. Then simplify by combining any like terms: $3x + 4 - 6\sqrt{3x + 2}$.

---

23. Find $(f \circ g)(x)$.

Solve It

24. Find $(g \circ f)(x)$.

Solve It

25. Find h(f(x)).

Solve It

26. Find $(f \circ f)(x)$.

Solve It

27. Find f(g(−1)).

Solve It

28. Find g(h(3)).

Solve It

Working Together: Domain and Range

Now that you've been combining and composing functions in the last two sections, you may be wondering what's happening with the domain and the range of the new function. Remember that domain is input, usually x, and that range is output, usually y. The truth is that the domain of the given function totally (like, for sure!) depends on the operation being performed and the original functions. It's possible that something changed, and it's also possible that nothing did. Typically, you'll be asked to find the domain of a combined function and *not* the range.

Pre-calc teachers and textbooks talk a lot about two functions whose domains are *not* all real numbers:

✔ **Rational functions:** The denominator of any fraction can't be 0, so it's possible that some rational functions are undefined because of this fact. Set the denominator equal to 0 and solve to find the restrictions on your domain.

✔ **Square root functions (or any even root):** The *radicand* (what's under the root sign) can't ever be negative. This affects domain; to find out how, set the radicand greater than or equal to 0 and solve. The solution to this inequality is your domain.

Undefined values are also called *excluded values*, so be on the lookout for your textbook to use that terminology as well. When you're asked to find the domain of a combined function, take your time. We can't put it into a nice, neat package and give you one rule that works all the time for finding a combined function's domain. Take a look at both of the original functions and ask yourself if their domains have any restrictions. These restrictions carry through and combine together to the new combined function.

You use those same three functions you've been using for the last two sections:

$$f(x) = x^2 - 6x + 2$$
$$g(x) = 2x^2 - 5x$$
$$h(x) = \sqrt{3x + 2}$$

Q. Find the domain of $f(h(x))$.

A. **The domain is all numbers greater than or equal to $-\frac{2}{3}$.** Take a look at the original two functions first. $f(x)$ is a polynomial; there are no restrictions on the domain. However, $h(x)$ is a square root function, so the radicand has to be positive. $3x + 2 \geq 0$; $3x \geq -2$; $x \geq -\frac{2}{3}$. The new combined function must honor this domain as well.

29. Find the domain of $(f \circ g)(x)$.

Solve It

30. Find the domain of $h(f(x))$.

Solve It

31. Find the domain of $(f + h)(x)$.

Solve It

32. Find the domain of $(h/g)(x)$.

Solve It

Finding the Inverse of a Function (Who Knew It Was Lost?)

An *inverse function* undoes what a function does. You've seen inverse operations before: Addition undoes subtraction and division undoes multiplication. It shouldn't surprise you, then, that functions have inverses. If $f(x)$ is the original function, then $f^{-1}(x)$ is the symbol for the inverse. This notation is only used for the inverse function and is *never* meant to represent $\dfrac{1}{f(x)}$.

You'll be asked to do three main things with inverses:

✔ Given a function, graph its inverse

✔ Find the inverse of a given function

✔ Show that two functions are inverses of each other

In any case, all you need to know is that it's all about input and output. If (a, b) is a point in the original function, then (b, a) is a point in the inverse function. Domain and range swap places from a function to its inverse. If asked to graph the inverse function, graph the original and then swap all x and y values in each point to graph the inverse. To find the inverse of a given function, literally take x and y (or $f(x)$) and switch them. After the swap, change the name to the symbol for an inverse function, $f^{-1}(x)$, and solve for the inverse. Lastly, to show that two functions $f(x)$ and $g(x)$ are inverses of each other, place one inside the other using composition of functions, $f(g(x))$, and simplify to show that you get x. Then do it the other way around with $g(f(x))$ to make sure it works both ways.

EXAMPLE

Q. Find the inverse of $f(x) = 5x - 4$.

A. $\mathbf{f^{-1}(x)} = \dfrac{x+4}{5}$. First, switch x and $f(x)$:

$x = 5f(x) - 4$. Name the new function by its correct name, the inverse function: $x = 5f^{-1}(x) - 4$. Now solve for the inverse: $x + 4 = 5f^{-1}(x)$; $\dfrac{x+4}{5} = f^{-1}(x)$.

Q. Determine whether $f(x) = 3x - 1$ and $g(x) = \dfrac{x+1}{3}$ are inverses of each other.

A. **These two functions are inverses.** First, find $(f \circ g)(x)$: $3(\dfrac{x+1}{3}) - 1$. Simplify this expression: $x + 1 - 1 = x$. That's what it's supposed to be, so move onto the next one: $(g \circ f)(x)$: $\dfrac{3x - 1 + 1}{3} = \dfrac{3x}{3} = x$. That one works, too, so these two functions are inverses of each other.

33. Graph the inverse of $g(x) = \sqrt{x-2}$.

Solve It

34. Find the inverse of $k(x) = \dfrac{3x}{x-1}$.

Solve It

35. Determine whether $f(x) = x^3 - 1$ and $g(x) = \sqrt[3]{x} + 1$ are inverses of each other.

Solve It

36. Determine whether $f(x) = \dfrac{1-x}{2}$ and $g(x) = 1 - 2x$ are inverses of each other.

Solve It

Answers to Questions on Functions

1 Is $f(x) = x^3 - 1$ even, odd, or neither? **The answer is neither.**

Find $f(-x) = (-x)^3 - 1 = -x^3 - 1$. This isn't the same function as the original, so the answer isn't even. It's also not the exact opposite of the original, so the answer isn't odd. The answer is that it's neither.

2 Determine whether the given graph is even, odd, or neither. **The answer is odd.**

If you look at the graph upside down, it looks exactly the same — that means it's odd.

3 Sketch half of the graph of $f(x) = \sqrt{x^2 - 4}$ and use symmetry to complete the graph. **See the graph for the answer.**

Find $f(-x)$ first and discover that the function doesn't change at all, which means you've got an even function. If you plug and chug some negative values for x, you know that the positive values for each corresponding x will be the same.

For example, $f(-2) = 0$, so you know that $f(2)$ is also 0. $f(-3) = \sqrt{5}$, and so does $f(3)$. $f(-5) = \sqrt{21}$, and so does $f(5)$. Knowing these points gives you the graph.

4 Sketch half the graph of $f(x) = 4x^3$ and use symmetry to complete the graph. **See the graph for the answer.**

If you find $f(-x)$, you get $-4x^3$, which is the exact opposite of the original function, meaning that you have an odd graph. Each x gives you a value $f(x)$, and each opposite $-x$ gives the opposite $-f(x)$.

Plug and chug some values to get the graph: $f(-3) = -108$, so $f(3) = 108$. $f(2) = 32$, so $f(-2) = -32$. $f(-1) = -4$, so $f(1) = 4$. Put these and as many other points as you'd like on the graph.

5 Graph the function $a(x) = -2(x - 1)^2$. **See the graph for the answer.**

This function takes the parent quadratic graph and moves it to the right by 1. The vertical transformation is 2, making each point twice as tall. The negative sign is a reflection, turning the graph upside down. Put all these pieces together to get the graph.

$a(x) = -2(x - 1)^2$

6 Graph the function $b(x) = |x + 4| - 1$. **See the graph for the answer.**

This absolute value function has a vertical shift of 4 to the left and a horizontal shift of 1 down. The coefficient in the front is 1, so it doesn't have a vertical transformation — the graph has just been moved.

$b(x) = |x + 4| - 1$

7 Graph the function $c(x) = \sqrt{x + 3}$. **See the graph for the answer.**

This square root function is shifted horizontally to the left by 3. Don't forget that those horizontal shifts are always the opposite of what they appear to be.

$c(x) = \sqrt{x + 3}$

8 Graph the function $f(x) = -x^2 - 6x$. **See the graph for the answer.**

This one doesn't look like any of the others that you've dealt with so far. That's kind of unfair of us, but it brings up the topic of conic sections, which we talk about later in depth in Chapter 12. A parabola is one of these fancy types of curves. To get this parabola into its graphable form, you have to follow a procedure known as *completing the square* (see Chapters 4 and 12 for more information on this procedure). We do it for you here and tell you that the function will become $f(x) = -1(x + 3)^2 + 9$. We include it here because a few textbooks (though not many) teach completing the square early so that you can graph these types of problems. If your teacher is mean enough to include one of these *without* teaching you how to complete the square, you have to plug and chug this type of problem in order to graph it — pick x-values to find the corresponding y-values. Just be sure that your final graph is a parabola, as is ours.

$f(x) = -x^2 - 6x$
$f(x) = -1(x + 3)^2 + 9$

9 $g(x) - 3$. **See the graph for the answer.**

This is just a graph that we made up. Because it's a function, it still follows all the rules of transforming functions. Take every single point on the given $g(x)$ function and shift each down by 3, and you end up with a graph that looks like ours.

$g(x) - 3$

10 $2g(x - 1)$. **See the graph for the answer.**

This time, $g(x)$ is shifted to the right by 1. Take the height of each point in the original function and multiply it by 2 to get the new height. For example, the original function passes through the point (2, 4). The height of this point is 4, so when you double that in the new graph, you make the height 8. Do this for every single point and end up with the graph.

$2g(x - 1)$

11 Graph the function $q(x) = \dfrac{-2}{(x - 4)(x + 5)}$. **See the graph for the answer.**

The vertical asymptote comes from the denominator: $(x - 4)(x + 5) = 0$. This equation is already neatly factored, so all you have to do is use the zero product property and set each factor equal to 0 and solve. If $x - 4 = 0$, then $x = 4$, and if $x + 5 = 0$, then $x = -5$. Put both of these on the graph as vertical asymptotes. The horizontal asymptote is the x-axis again because the denominator has the greater degree. The intervals you need to take a closer look at are $(-\infty, -5)$, $(-5, 4)$, and $(4, \infty)$.

Pick a couple of x-values from each interval to get an idea of what the graph is doing. When $x = -7$, $y = -0.09$, and when $x = -6$, $y = -0.2$ — both below the horizontal asymptote. When $x = -4$, $y = 0.25$; when $x = -1$, $y = 0.1$; when $x = 1$, $y = 0.11$; and when $x = 3$, $y = 0.25$. These are all above the horizontal asymptote. On the final interval, when $x = 5$, $y = -0.2$, and when $x = 6$, $y = -0.09$. These are both below the horizontal asymptote. Put all the pieces together in the final graph.

$$q(x) = \frac{-2}{(x-4)(x+5)}$$

12 Graph the function $r(x) = \frac{x+3}{x^2-x-6}$. **See the graph for the answer.**

First up, vertical asymptote(s). Set $x^2 - x - 6 = 0$ and factor to $(x-3)(x+2) = 0$. Set each factor equal to 0 and solve. If $x - 3 = 0$, then $x = 3$, and if $x + 2 = 0$, then $x = -2$. Add these two vertical asymptotes to your graph. Next up is the horizontal asymptote. Because the denominator has the greater degree, the horizontal asymptote is the x-axis again. Notice, however, that now that the variable is in the numerator as well, there may be an x-intercept. Set the numerator equal to 0 and solve. $x + 3 = 0$ tells you that $x = -3$ is an intercept. The graph crosses the x-axis even though it isn't supposed to. Use this fact to set up the intervals: $(-\infty, -3)$, $(-3, -2)$, $(-2, 3)$, and $(3, \infty)$. Each interval is, respectively, below, above, below, and above the horizontal asymptote. The graph looks a little weird, but then, which of these problems *doesn't* look weird?

$$r(x) = \frac{x+3}{x^2-x-6}$$

13 Graph the function $t(x) = \frac{x^2-5x}{x^2-4x-21}$. **See the graph for the answer.**

Find the vertical asymptotes for this one by factoring the denominator. If $x^2 - 4x - 21 = 0$, then $(x-7)(x+3) = 0$. This gives you two solutions: $x = 7$ and $x = -3$. The degrees are the same again, so the horizontal asymptote this time is $y = 1$. Put the asymptotes onto the graph and then pick x-values to get the graph.

14 Graph the function $u(x) = \dfrac{x^2 - 10x - 24}{x + 1}$. **See the graph for the answer.**

This is the toughest of all the problems here because there's no horizontal asymptote — the numerator has the greater degree. Use long division to find the quotient $x - 11$; graph this as an equation, $y = x - 11$, with a dotted line to mark your oblique asymptote. Next, add the vertical asymptote by solving the equation $x + 1 = 0$ to get $x = -1$. Finally, plug and chug some values on each interval to get the graph.

15 Graph $g(x) = \begin{cases} \sqrt{x + 3} & \text{if } x \le -1 \\ (x + 3)^2 & \text{if } x > -1 \end{cases}$. **See the graph for the answer.**

And now you get to graph piece-wise functions together. Take a look at each interval of the domain to determine the graph's shape. For this function, the top piece is only defined when $x \le -1$. This part of the graph looks like a square root graph shifted 3 to the left. The bottom piece is defined when $x > -1$. This part of the graph is a parabola, shifted 3 to the left. If it helps you to lightly sketch the whole graph and then erase the part you don't need, we highly recommend it.

$$\begin{cases} \sqrt{x+3} \text{ if } x \le -1 \\ (x+3)^2 \text{ if } x > -1 \end{cases}$$

16 Graph $h(x) = \begin{cases} \dfrac{1}{2}x - 4 \text{ if } x \le -2 \\ 3x + 3 \text{ if } -2 < x < 2 \\ 4 - x \text{ if } x \ge 2 \end{cases}$. **See the graph for the answer.**

The first piece is a linear function that's defined only when $x \le -2$. The second piece is also a linear function, defined between -2 and 2. The third piece is another linear function, defined when $x \ge 2$.

$$\begin{cases} \frac{1}{2}x - 4 \text{ if } x \le 2 \\ 3x + 3 \text{ if } -2 < x < 2 \\ 4 - x \text{ if } x \ge 2 \end{cases}$$

17 Graph $m(x) = \begin{cases} x^3 + 2 \text{ if } x < 0 \\ x^2 + 2 \text{ if } 0 \le x < 2 \\ x + 2 \text{ if } x \ge 2 \end{cases}$. **See the graph for the answer.**

The first piece is a cubic shifted up by 2 — its right endpoint should be open. However, when you graph the second piece, it's a parabola that's shifted up by 2. Its left endpoint overlaps the right endpoint of the first piece. This fills the hole that was there, and the graph carries on until $x = 2$, where it gets broken again. The third piece follows the linear function to the right of $x = 2$. The graph comes together to look like this:

18 Graph $n(x) = \begin{cases} |x-1| \text{ if } x < -3 \\ -3 \text{ if } x = 3 \\ |x| - 1 \text{ if } x > -3 \end{cases}$. **See the previous graph for the answer.**

This piece-wise function is different because the middle piece is defined only at one point. When $x = 3$, $y = -3$; that's it. The first piece follows the absolute value graph that has been shifted to the right by 1. The third piece is also an absolute value graph, but it has been shifted down 1. Here's the not-really-last graph.

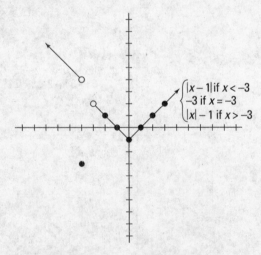

19 Find $(f + h)(x)$. **The answer is $x^2 - 6x + 2 + \sqrt{3x+2}$.**

Take the f function and add the h function to it. Because one is a polynomial and the other is a square root, there are no pesky like terms. The answer is $(f + h)(x) = x^2 - 6x + 2 + \sqrt{3x+2}$.

20 Find $(fg)(x)$. **The answer is $2x^4 - 17x^3 + 34x - 10x$.**

Start off by writing out what you've been asked to find — the product of f and g: $(fg)(x) = (x^2 - 6x + 2)(2x^2 - 5x)$. Distribute each term of the left polynomial to each term of the right polynomial to get $2x^4 - 12x^3 + 4x^2 - 5x^3 + 30x^2 - 10x$. Next, combine the like terms and get $(fg)(x) = 2x^4 - 17x^3 + 34x^2 - 10x$.

21 Find (h/g)(x). Are there any undefined values for this new function? **The answer is** $\dfrac{\sqrt{3x+2}}{2x^2-5x}$ **and yes, there are undefined values:** $x = 0$ **and** $x = \frac{5}{2}$.

This time you're asked to find the quotient of h and g, with h on the top and g on the bottom. Because they're different types of functions, they won't simplify at all. This means your answer is, simply, $\dfrac{\sqrt{3x+2}}{2x^2-5x}$. Because the denominator now has a polynomial, there may be undefined values. Set the denominator equal to 0 to start: $2x^2 - 5x = 0$. Solve by factoring out the GCF: $x(2x-5)$. This has two solutions: $x = 0$ and $x = \frac{5}{2}$. These are the undefined values.

22 Find (g + h)(2). **The answer is** $2x^2 - 5x + \sqrt{3x+2}$.

Because you're adding another polynomial and a square root, there are no like terms. The answer is $2x^2 - 5x + \sqrt{3x+2}$.

23 Find $\left(f \circ g\right)\left(x\right)$. **The answer is** $4x^4 - 20x^3 + 13x^2 + 30x + 2$.

Take the g function and start plugging into f everywhere it says x: $(2x^2 - 5x)^2 - 6(2x^2 - 5x) + 2$. Multiply everything out first: $4x^4 - 20x^3 + 25x^2 - 12x^2 + 30x + 2$. Now combine like terms to get the answer: $\left(f \circ g\right)\left(x\right) = 4x^4 - 20x^3 + 13x^2 + 30x + 2$.

24 Find $\left(g \circ f\right)\left(x\right)$. **The answer is** $2x^4 - 24x^3 + 75x^2 - 18x - 2$.

This time, place f into g where it says x: $2(x^2 - 6x + 2)^2 - 5(x^2 - 6x + 2)$. Square the polynomial on the left first by multiplying $x^2 - 6x + 2$ by itself and distributing each term by each term. This is quite long, so we show you the steps so you can follow along. $2(x^4 - 6x^3 + 2x^2 - 6x^3 + 36x^2 - 12x + 2x^2 - 12x + 4) - 5(x^2 - 6x + 2)$. Combine like terms: $2(x^4 - 12x^3 + 40x^2 - 24x + 4) - 5(x^2 - 6x + 2)$. Distribute the coefficients next: $2x^4 - 24x^3 + 80x^2 - 48x + 8 - 5x^2 + 30x - 10$. Combine the like terms to end up with the final answer: $\left(g \circ f\right)\left(x\right) = 2x^4 - 24x^3 + 75x^2 - 18x - 2$.

25 Find h(f(x)). **The answer is** $\sqrt{3x^2 - 18x + 8}$.

Substitute f in for x in the h function: $\sqrt{3\left(x^2 - 6x + 2\right) + 2}$. Distribute that 3 inside the root: $\sqrt{3x^2 - 18x + 6 + 2}$. Combine those like terms to end up with the answer: h(f(x)) = $\sqrt{3x^2 - 18x + 8}$.

26 Find $\left(f \circ f\right)\left(x\right)$. **The answer is** $x^4 - 12x^3 + 34x^2 + 12x - 6$.

This time, place f into itself everywhere it says x: $(x^2 - 6x + 2)^2 - 6(x^2 - 6x + 2) + 2$. You went through the process of squaring that polynomial once before in question 24, so we won't do it again here. When you multiply everything out, you get $x^4 - 12x^3 + 40x^2 - 24x + 4 - 6x^2 + 36x - 12 + 2$. Combine like terms to end up with the answer: $\left(f \circ f\right)\left(x\right) = x^4 - 12x^3 + 34x^2 + 12x - 6$.

27 Find f(g(−1)). **The answer is 9.**

You already found f(g(x)) in question 23 — it's $4x^4 - 20x^3 + 13x^2 + 30x + 2$. Now, substitute −1 in for x: $4(-1)^4 - 20(-1)^3 + 13(-1)^2 + 30(-1) + 2$. Simplify by dealing with all the exponents first: $4(1) - 20(-1) + 13(1) + 30(-1) + 2$. Simplify further by multiplying: $4 + 20 + 13 - 30 + 2$. Add and subtract to finally end up with 9.

28 Find g(h(3)). **The answer is** $22 - 5\sqrt{11}$.

You didn't find g(h(x)) in any other problem, but that doesn't mean you have to. Remember that these are read right to left. This question is asking you to plug 3 into h and then plug that answer into g. Start with h(3) = $\sqrt{3(3) + 2} = \sqrt{9+2} = \sqrt{11}$. Now plug this value in g and find $g\left(\sqrt{11}\right)$: $2\left(\sqrt{11}\right)^2 - 5\sqrt{11} = 2(11) - 5\sqrt{11} = 22 - 5\sqrt{11}$.

29 Find the domain of $(f \circ g)(x)$. **The domain is all real numbers.**

You found the composition of $(f \circ g)(x)$ in question 23, and you ended up with a longish polynomial: $(f \circ g)(x) = 4x^4 - 20x^3 + 13x^2 + 30x + 2$. Because it's a polynomial, there's nothing weird about the domain. It's all real numbers.

30 Find the domain of $h(f(x))$. **The domain is $x < 0.48$ and $x > 5.52$.**

$h(f(x)) = \sqrt{3x^2 - 18x + 8}$, which you found in question 25. This puts a polynomial under a square root. A square root's radicands have to be positive. Find where $3x^2 - 18x + 8$ is positive by setting it greater than or equal to 0 and solving using the quadratic formula. This gives you two critical values: $\dfrac{18 \pm \sqrt{228}}{6}$. Place these test values on a number line and look at the intervals that are determined by them: $x < 0.48$, $0.48 < x < 5.52$, and $x > 5.52$. If you plug test values from each interval into the inequality, you discover which intervals work and which don't. In this case, x has to be less than 0.48 and bigger than 5.52, thus the answer $x < 0.48$ and $x > 5.52$.

31 Find the domain of $(f + h)(x)$. **The domain is $x \geq -\frac{2}{3}$.**

$(f + h)(x) = x^2 - 6x + 2 + \sqrt{3x + 2}$. This adds a square root to a polynomial, so the new combined function must follow all the rules that the square root function did by itself.

32 Find the domain of $(h/g)(x)$. **The domain is $x \geq -\frac{2}{3}$, except $x = 0$ and $x = \frac{5}{2}$.**

$(h/g)(x) = \dfrac{\sqrt{3x + 2}}{2x^2 - 5x}$. The square root in the numerator restricts the domain to $x \geq -\frac{2}{3}$. The polynomial in the denominator has undefined values $x = 0$ and $x = \frac{5}{2}$. These are both in the restricted domain, so they become part of the answer. You express it as one neat sentence: "The domain is $x \geq -\frac{2}{3}$, except $x = 0$ and $x = \frac{5}{2}$."

33 Graph the inverse of $g(x) = \sqrt{(x - 2)}$. **See the previous graph for the answer.**

No more graphing! We mean it this time . . . in this chapter anyway. Start off by graphing the square root function shifted to the right by 2. Points on this graph include (2, 0), (3, 1), and (6, 2). Flip them to get (0, 2), (1, 3), and (2, 6) — all points on the inverse function graph. We show both in the graph.

34 Find the inverse of $k(x) = \dfrac{3x}{x-1}$. **The answer is $f^{-1}(x) = \dfrac{x}{x-3}$.**

Switch x and $f(x)$ and name the new inverse by its real name: $x = \dfrac{3f^{-1}(x)}{f^{-1}(x)-1}$. Notice the inverse is in the numerator and the denominator. The only way you can solve for it is to get rid of the fraction first by multiplying both sides by the denominator and getting $x(f^{-1}(x) - 1) = 3f^{-1}(x)$. Distribute the x and get $x f^{-1}(x) - x = 3 f^{-1}(x)$. Get everything with the inverse in it to one side and everything else to the other side: $xf^{-1}(x) - 3f^{-1}(x) = x$. Factor out the GCF on the left, the inverse function: $f^{-1}(x)(x-3) = x$. Now divide the leftovers to solve for the inverse: $f^{-1}(x) = \dfrac{x}{x-3}$.

35 Determine whether $f(x) = x^3 - 1$ and $g(x) = \sqrt[3]{x} + 1$ are inverses of each other. **The answer is that they're not inverses.**

First, find the composition $(f \circ g)(x)$. $(f \circ g)(x) = \left(\sqrt[3]{x} + 1\right)^3 - 1$. This doesn't simplify to get x, so you can stop. They're not inverses.

36 Determine whether $f(x) = \dfrac{1-x}{2}$ and $g(x) = 1 - 2x$ are inverses of each other. **The answer is yes, they are inverses.**

$(f \circ g)(x) = \dfrac{1-(1-2x)}{2} = \dfrac{1-1+2x}{2} = \dfrac{2x}{2} = x$. One down, one to go.

$(g \circ f)(x) = 1 - 2\left(\dfrac{1-x}{2}\right) = 1 - (1-x) = 1 - 1 + x = x$. Okay, you checked both; it's official, they're inverses.

Chapter 4

Go Back to Your Roots to Get Your Degree

A polynomial is any expression with more than one term in it. The highest exponent on any term in a polynomial is its *degree*. In this chapter, we review solving polynomial equations to find the solutions, which are also called *roots* or *zeros*. We start with a review of solving *quadratic equations* — polynomials where the highest exponent is two. Then we move into equations with higher degrees and show you how to solve them. We also take a look at using roots to factor polynomials and how to graph polynomials.

Reason Through It: Factoring a Factorable Polynomial

Before getting started on the nitty-gritty, here's some vocabulary you should know to be successful in this chapter (and after):

Standard form: What most textbooks use to write a quadratic equation: $ax^2 + bx + c = 0$

Quadratic term: The term with the second degree: ax^2

Linear term: The term with the linear degree: bx

Constant: The term with zero degree: c

Leading coefficient: The number multiplying the term with the highest degree: a

In math, the process of breaking down a polynomial into the product of two polynomials with a smaller degree is called *factoring*. In general, factoring works best on quadratic equations and is **always** the first thing you should try when asked to solve second-degree polynomials. Some types of factoring (like the difference of cubes or grouping — more on those later in this section) may work on higher degree polynomials, and you should always check them to see if they're factorable first. When presented with a polynomial and asked to solve it, you should always try the following methods of factoring, in order:

- ✔ **Greatest common factor:** The greatest common factor, or GCF, is the biggest expression that will divide into all the other terms. It's a little like doing the distributive property backwards.

  Break each term down into prime factors, look at all those factors to see what they share in common (that's your GCF), factor the GCF out from every term by putting it in front of a set of parentheses, and leave the factors that aren't the GCF inside the parentheses.

- ✔ **The polynomial is a binomial:** If the polynomial has two terms, check to see whether it's a difference of squares or the sum or difference of cubes.

  Difference of squares $a^2 - b^2$ always factors to $(a - b)(a + b)$

  Difference of cubes $a^3 - b^3$ always factors to $(a - b)(a^2 + ab + b^2)$

  Sum of cubes $a^3 + b^3$ always factors to $(a + b)(a^2 - ab + b^2)$

- ✔ **The polynomial is a trinomial:** Try using the FOIL method backwards.

  Some teachers teach the "guess and check method," where you keep trying different pairs of binomials until you happen to stumble on the right one. This isn't fun by any means, and you could try all day long and never figure it out (or maybe the polynomial is prime and won't factor).

  We recommend using the *British method* (also known as the *FOIL method backwards*) instead. Follow these steps to use this method:

  1. **Multiply the quadratic term and the constant term.** You only do this in your head (or somewhere else on your paper) and you only do it to proceed to the next step.

  2. **Write down all the factors of the result of Step 1, in pairs.** Again, you do this for you only and also to make sure that you list every possibility — that's why it's not guess and check. If you list them all and none of them work (see Step 3), you know your trinomial is prime.

  3. **Find the pair from the list in Step 2 that adds to produce the linear term.** Only one of them will work, and if none of them do, it's prime.

  4. **Break up the linear term into two terms — the winning pair from Step 3.**

  You've now created a polynomial with four terms. Proceed to the next type of factoring — a polynomial with more than three terms.

- ✔ **The polynomial has more than three terms:** Try grouping the polynomial.

  Group the polynomial into two sets of two. Find the GCF for each set and factor it out. Find the GCF of the two remaining expressions and factor it out. You end up with two binomials, exactly what you were looking for!

After you have the polynomial factored, you can use the zero product property to solve it by setting each factor equal to 0 and solving.

Q. Solve the equation $3x^2 + x - 2 = 0$.

A. **$x = \frac{2}{3}$, $x = -1$.** Multiply the quadratic term and the constant term: $(3x^2)(-2) = -6x^2$. Write down all factors of this, in pairs: $-x$ and $6x$, x and $-6x$, $-2x$ and $3x$, $2x$ and $-3x$. The pair that adds up to the linear term is $-2x$ and $3x$. Split the middle term into two using this pair: $3x^2 - 2x + 3x - 2 = 0$. Now that you have four terms instead of three, use grouping to factor it: $x(3x - 2) + 1(3x - 2) = 0$. Notice that the second two factors only have a GCF of one; you still factor it out. Now there's a GCF again — both sets of terms share $(3x - 2)$ so that can factor out to the front: $(3x - 2)(x + 1) = 0$. Finally, use the zero product property to solve the equation. If $3x - 2 = 0$, then $x = \frac{2}{3}$; and if $x + 1 = 0$, then $x = -1$.

Q. Solve the equation $3x^3 - 3x = 0$.

A. **$x = 0$, 1, and -1.** Always check for the GCF first and factor it out: $3x(x^2 - 1) = 0$. Now recognize the "leftovers" as a difference of squares which factors again: $3x(x - 1)(x + 1) = 0$. Set each factor equal to 0 and solve: $3x = 0$, $x = 0$. $x - 1 = 0$, $x = 1$. $x + 1 = 0$, $x = -1$.

1. Solve the equation $2y^2 + 5y = 12$.

Solve It

2. Solve the equation $16m^2 - 8m + 1 = 0$.

Solve It

3. Solve the equation $x^3 + x^2 = 9x + 9$.

Solve It

4. Solve the equation $\frac{1}{6}x^2 + \frac{2}{3}x = 2$.

Solve It

Get Your Roots Done while Solving a Quadratic Polynomial

What happens when a quadratic equation doesn't factor? You're done, right? Well, not quite. You have two more methods you can use. One we always recommend (the quadratic formula) and the other we don't (completing the square). However, your teacher may require you to use both, so we include both here. When you get to graph quadratics in Chapters 3 and 12, it's easiest to do by completing the square and then using the rules of transforming a parent function to get the graph.

Completing the square

Completing the square is the technique to use when you're specifically told to do so. Other than that, save it for graphing. Here are the steps:

1. **Make sure the quadratic is written in standard form:** $ax^2 + bx + c = 0$.

2. **Add (or subtract) the constant term from both sides:** $ax^2 + bx = -c$.

3. **Factor out the leading coefficient from the quadratic term and the linear term:** $a\left(x^2 + \dfrac{b}{a}x\right) = -c$.

4. **Divide the new linear coefficient by two:** $\dfrac{b}{a} \div 2 = \dfrac{b}{2a}$; **square this:** $\left(\dfrac{b}{2a}\right)^2 = \dfrac{b^2}{4a^2}$; **and add this inside the parentheses:** $a\left(x^2 + \dfrac{b}{a}x + \dfrac{b^2}{4a^2}\right) = -c$.

5. **Keep the equation balanced by multiplying the leading coefficient by the term you just added in Step 4:** $a \cdot \dfrac{b^2}{4a^2} = \dfrac{b^2}{4a}$, **and adding it to the other side:** $a\left(x^2 + \dfrac{b}{a}x + \dfrac{b^2}{4a^2}\right) = -c + \dfrac{b^2}{4a}$.

6. **Divide the leading coefficient from both sides:** $x^2 + \dfrac{b}{a}x + \dfrac{b^2}{4a^2} = -\dfrac{c}{a} + \dfrac{b^2}{4a^2}$.

7. **Factor the trinomial on the left side of the equation:** $\left(x + \dfrac{b}{a}\right)^2 = -\dfrac{c}{a} + \dfrac{b^2}{4a^2}$.

8. **Take the square root of both sides:** $x + \dfrac{b}{a} = \pm\sqrt{-\dfrac{c}{a} + \dfrac{b^2}{4a^2}}$.

9. **Solve for x:** $x = \dfrac{-b}{a} \pm \sqrt{-\dfrac{c}{a} + \dfrac{b^2}{4a^2}}$.

Quadratic formula

Of course, those of you who know the quadratic formula should vaguely recognize the steps above — they're the derivation of the quadratic formula. All you have to do is find the common denominator of both the fractions inside the square root, add them together, and watch the square root simplify. Ultimately, you end up with the quadratic formula:

$$x = \frac{-b \pm \sqrt{b^2 - 4ac}}{2a}$$

EXAMPLE

Q. Solve the equation $5x^2 - 12x - 2 = 0$.

A. $x = \dfrac{6 \pm \sqrt{46}}{5}$. This equation doesn't factor, so you use the quadratic formula to solve it. $a = 5$, $b = -12$, $c = -2$. Plug these values into the quadratic formula:

$\dfrac{-(-12) \pm \sqrt{(-12)^2 - 4(5)(-2)}}{2(5)}$. Now simplify it:

$\dfrac{12 \pm \sqrt{144 + 40}}{10} = \dfrac{12 \pm \sqrt{184}}{10}$. Don't forget to check your square roots and simplify them as well: $\dfrac{12 \pm 2\sqrt{46}}{10}$. Finally, 2 goes into every coefficient and constant in the answer, so it simplifies even further to $\dfrac{6 \pm \sqrt{46}}{5}$.

---

5. Solve $x^2 - 10 = 2x$.

Solve It

6. Solve $7x^2 - x + 2 = 0$.

Solve It

---

7. Solve $x^2 - 4x - 7 = 0$ by completing the square.

Solve It

8. Solve $-2.31x^2 - 4.2x + 6.7 = 0$.

Solve It

Climb the Mountains by Solving High Order Polynomials

The greater the degree of your given polynomial, the harder it is to solve the equation by factoring. You should always still try that first because you never know . . . it may actually work! When factoring fails, however, you begin anew with a longer and more complicated process for finding the roots. We walk you through each step, one by one.

Always begin by finding the degree of the polynomial because it gives you some very important information about your graph. The degree of the polynomial tells you the greatest maximum number of roots — it's that easy. A fourth-degree polynomial will have up to, but no more than, four roots.

Determining positive and negative roots: Descartes' Rule of Signs

When you know the total number of roots, you can use Descartes' Rule of Signs to determine how many of the roots are positive and how many are negative. This literally tells you how many times your graph crosses the x-axis on the negative side as well as on the positive side. These will all be real roots represented as points on the real number line that is the x-axis. All you have to be able to do is count!

Make sure that the polynomial $f(x)$ is written in descending order first, from highest degree to lowest. Look at the sign of each term and count how many times the sign changes from positive to negative and vice versa. The number of sign changes represents the maximum number of positive real roots. The rule also says that this number decreases by 2 over and over again until you end up with 1 or 0 (more on this in the section on imaginary roots). This gives you the list of the possible number of real positive roots.

Descartes also figured out that if you take a look at $f(-x)$ and count again, you discover the maximum number of negative real roots. Remember that negative numbers raised to even powers are positive, and negative numbers raised to odd powers are negative. This means that $f(-x)$ changes from $f(x)$ only on the odd degrees. Each odd exponent becomes the opposite of what it was in $f(x)$. Count the number of times the sign changes in this function, subtract 2 over and over until you end up at 1 or 0, and end up with a list of the possible number of real negative roots.

Counting on imaginary roots

Imaginary roots happen in a quadratic equation when the radicand is negative. Remember from Algebra II that you should look at the *discriminant* (the part of the quadratic formula under the root sign: $b^2 - 4ac$). If the discriminant is negative, the roots are imaginary. The $\pm$ sign also tells you that there are two of these roots, always in pairs. This is why in the previous section you subtract by 2; you have to account for the fact that the roots may be in pairs of imaginary numbers. In fact, the pairs will always be *complex conjugates* of each other — if one root is $a + bi$, for example, then the other one is $a - bi$.

The *Fundamental Theorem of Algebra* says that every polynomial has at least one root in the complex number system. Chapter 11 explains complex numbers in depth, but for now all you need to know is that a complex number has both a real and an imaginary part. That's why there has to be at least one complex root — it has to be either real, imaginary, or both at the same time. Kinda cool, huh?

Because you know the total number of possible roots and the list of possible positive and negative roots, you can use all that information to determine how many imaginary roots a polynomial has. Pair up every possible number of positive roots with every possible number of negative roots. The remaining number of roots in each situation represents the imaginary roots.

Getting the rational roots

The *Rational Root Theorem* helps you narrow down the possibilities even further. Right now, if you've gone through all the steps, you only know the total number of roots, how many are positive real, how many are negative real, and how many are imaginary. That still leaves an infinite number of possibilities! The Rational Root Theorem helps you because it finds the possible roots that are *rational* (those that can be written as a fraction). The problem with the theorem? Not all roots are rational. Keep in mind that some (or all) of the roots are irrational.

To use the Rational Root Theorem, take all the factors of the constant term and divide by all the factors of the leading coefficient. This produces a list of fractions that are all possibilities for roots. You *could* try plugging each one of these possibilities into the original function in the hopes of finding a root (remember, they're also called zeros because the value of the function will be 0). This process is long and tedious because, each time, you're dealing with the original function. If there are 50 roots, you're not helping yourself by plugging and chugging. Instead, move on to the next step.

Synthetic division finds some roots

With the list from the last section in front of you, pick one fraction and try it to see if it works. If it does, the quotient is a *depressed polynomial*. No, it's not sad — its degree will be less than the one you started with. You use this quotient to find the next one, each time lessening the degree, which narrows down the roots you have to find. At some point, your polynomial will end up as a quadratic equation, which you can solve using factoring or the quadratic formula. Now that's clever! If the root you try doesn't work, you should always try it again to see if it's a root with *multiplicity* — that is, roots that are used more than once.

Here are the steps to use for synthetic division:

1. **Make sure the polynomial is written in descending order. If any degrees are missing, fill in the gaps with zeros.**

2. **Write the number that's the root you're testing outside the synthetic division sign. Write the coefficients of the polynomial in descending order and include any zeros from Step 1 inside the synthetic division sign.**

3. **Drop the first coefficient down.**

4. **Multiply the root on the outside and this coefficient. Write this product above the synthetic division line.**

5. **Add the next coefficient and the product from Step 4. This answer goes below the line.**

6. **Multiply the root on the outside and the answer from Step 5.**

7. **Repeat over and over again until you use all the coefficients.**

This process is easier to see with an example. Hold on and we'll show you. Just know that when you do synthetic division, you end up with a list of roots that actually work in the polynomial.

Q. Find the roots of the equation $x^3 + x^2 - 5x + 3 = 0$.

A. **$x = 1$ (double root), $x = -3$.** We go through the whole process described in this section for this example question.

The number of roots: First, this equation is third degree, so it may have up to three different roots.

Descartes' Rule of Signs: Next, by looking at $f(x) = x^3 + x^2 - 5x + 3$, you notice that the sign changes twice (between the second and third terms and the third and fourth terms). This means there could be two or zero positive real roots. Next, look at $f(-x) = -x^3 + x^2 + 5x + 3$ and notice the sign only changes once, giving you only one negative real root.

Imaginary roots: So if two roots are positive and one is negative, that leaves none leftover that are imaginary. But if zero are positive and one is negative, that leaves two imaginary roots.

Rational Root Theorem: Take all the factors of 3 (the constant term) and divide by all the factors of 1 (the leading coefficient) to determine the possible rational roots — $\pm \frac{1}{1}, \pm \frac{3}{1}$. Reduce the fractions and discard any duplicates to get the final list: $\pm 1, \pm 3$.

Synthetic division: Pick a root, any root, and use synthetic division to test and see if it actually is a root. Because we know the answers (we *did* write the question), we have you start with $x = 1$:

$$
\begin{array}{r|rrrr}
1 & 1 & 1 & -5 & -3 \\
  &   & 1 & 2 & -3 \\
\hline
  & 1 & 2 & -3 & 0
\end{array}
$$

The last column on the right is the remainder; because it's 0, you know you have one root: $x = 1$. Also notice that the other numbers are the coefficients of the depressed polynomial you're now working with: $x^2 + 2x - 3 = 0$. Because this is a quadratic, we recommend shifting gears and factoring it to $(x + 3)(x - 1) = 0$ to be able to use the zero product property to solve and get $x = -3$ and $x = 1$ (again — making it a double root!).

Q. Solve the equation $x^3 + 8x^2 + 22x + 20 = 0$.

A. **$x = -2, x = -3 \pm i$.** This equation is also a third degree, so it will have a maximum of three roots. Looking at $f(x) = x^3 + 8x^2 + 22x + 20$ reveals that none of them are positive. Looking at $f(-x) = -x^3 + 8x^2 - 22x + 20$ reveals that either three or one of them are negative. If zero are positive and three are negative, there can't be any imaginary roots. However, if zero are positive and only one is negative, two of them have to be imaginary. The Rational Root Theorem generates this list of fractions (and we're only looking at the negatives because we know there aren't any positive roots): $-\frac{1}{1}, -\frac{2}{1}, -\frac{4}{1}, -\frac{5}{1}, -\frac{10}{1}$ and $-\frac{20}{1}$. These all reduce, respectively, to $-1, -2, -4, -5, -10$, and -20. Start off with $x = -2$ to discover one of your roots:

$$
\begin{array}{r|rrrr}
-2 & 1 & 8 & 22 & 20 \\
   &   & -2 & -12 & -20 \\
\hline
   & 1 & 6 & 10 & 0
\end{array}
$$

The reduced polynomial you're now working with is $x^2 + 6x + 10$. This quadratic doesn't factor, so you use the quadratic formula to find that the last two roots are indeed imaginary: $x = -3 \pm i$.

9. Solve the equation $2x^3 + 3x^2 - 18x + 8 = 0$.

Solve It

10. Solve the equation $12x^4 + 13x^3 - 20x^2 + 4x = 0$.

Solve It

11. Solve the equation $x^3 + 7x^2 + 13x + 4 = 0$.

Solve It

12. Find the roots of the equation $x^4 + 10x^3 + 38x^2 + 66x + 45$.

Solve It

Strike That! Reverse It! Using Roots to Find an Equation

The *factor theorem* says that if you know the root of a polynomial, then you also know a factor of the polynomial. These two go back and forth, one to the other — roots and factors are interchangeable. Your textbook may ask you to factor a polynomial with a degree higher than two, and it just won't factor using any of the techniques we describe in the earlier sections. In this case, you must find the roots and use them to find the factors.

TIP

If $x = c$ is a root, then $x - c$ is a factor and vice versa. It always works, and that's something you can count on. Nice, huh?

EXAMPLE

Q. Use the roots of $x^3 + x^2 - 5x + 3 = 0$ to factor the equation.

A. $(x - 1)^2(x + 3) = 0$. This is the question from the first example in the last section. You found that the roots are $x = 1$ (double root) and $x = -3$. Using the factor theorem, if $x = 1$ is a root, then $x - 1$ is a factor (twice); and if $x = -3$ is a root, then $x - (-3)$, or $x + 3$, is a factor. This means that $x^3 + x^2 - 5x + 3 = 0$ factors to $(x - 1)^2(x + 3) = 0$.

---

13. If the roots of a polynomial are $x = -3, -2, 4$, and 6, what's the polynomial?

Solve It

14. If the roots of a polynomial are $x = 2$ and $4 \pm 3i$, what's the polynomial?

Solve It

---

15. Factor the polynomial $6x^4 - 7x^3 - 18x^2 + 13x + 6 = 0$.

Solve It

16. Factor the polynomial $x^4 + 10x^3 + 38x^2 + 66x + 45$.

Solve It

Graphing Polynomials

Now that you have your list of the roots of your polynomial, you've done the hard work to graph the polynomial. Remember that roots or zeros are x-intercepts — you now know where the graph crosses the x-axis. Follow these steps to get to the graph:

1. **Mark the x-intercepts on your graph.**

2. **Find the y-intercept by letting $x = 0$.** The shortcut? It will *always* be the constant term.

3. Use the *leading coefficient test* to determine which of the four possible ways the ends of your graph will point:

   a. If the degree of the polynomial is even and the leading coefficient is positive, both ends of the graph will point up.

   b. If the degree of the polynomial is even and the leading coefficient is negative, both ends of the graph will point down.

   c. If the degree of the polynomial is odd and the leading coefficient is positive, the left side of the graph will point down and the right side will point up.

   d. If the degree of the polynomial is odd and the leading coefficient is negative, the left side of the graph will point up and the right side will point down.

4. **Figure out what happens in between the *x*-intercepts by picking any *x*-value on each interval and plugging it into the function to determine if it's positive (and, therefore, above the *x*-axis) or negative (below the *x*-axis).**

5. **Plot the graph by using all the information you've determined.**

Q. Graph the equation $f(x) = x^3 + x^2 - 5x + 3$.

A. **See the graph in Figure 4-1.** This is the first example from the section on solving higher order polynomials again. You found that the roots are $x = 1$ (double root) and $x = -3$. The y-intercept is the constant $y = 3$. The leading coefficient test tells you the graph starts by pointing down and ends by pointing up. The double root at $x = 1$ makes the graph "bounce" and not cross there.

Figure 4-1:
The graph of
$f(x) = x^3 + x^2$
$- 5x + 3$.

EXAMPLE

Q. Graph the equation $f(x) = x^3 + 8x^2 + 22x + 20$.

A. **See the graph in Figure 4-2.** This is the second example from the section on solving higher order polynomials. You found one real root of $x = -2$, as well as the complex conjugates $x = -3 \pm i$.

Figure 4-2:
The graph of
$f(x) = x^3 + 8x^2$
$+ 22x + 20$.

17. Graph $f(x) = x^4 + 2x^3 - 13x^2 - 14x + 24$.

Solve It

18. Graph $f(x) = 6x^4 - 7x^3 - 18x^2 + 13x + 6$.

Solve It

19. Graph $f(x) = 12x^4 + 13x^3 - 20x^2 + 4x$.

Solve It

20. Graph $f(x) = x^4 + 10x^3 + 38x^2 + 66x + 45$.

Solve It

Answers to Questions on Finding Roots

1. Solve the equation $2y^2 + 5y = 12$. **The answer is $y = \frac{3}{2}, -4$.**

Begin with any quadratic equation by getting 0 on one side of the equation. In this case, subtract 12 from both sides: $2y^2 + 5y - 12 = 0$. Now, begin factoring by multiplying the leading term ($2y^2$) and the constant term (-12) to get $-24y^2$. List all the factors of this, in pairs: $-y$ and $24y$, y and $-24y$, $-2y$ and $12y$, $2y$ and $-12y$, $-3y$ and $8y$, $3y$ and $-8y$, $-4y$ and $6y$, $4y$ and $-6y$. The correct pair that adds up to the linear term is $-3y$ and $8y$. Split up the trinomial into a polynomial using this magic pair: $2y^2 - 3y + 8y - 12 = 0$. Factor by grouping: $y(2y - 3) + 4(2y - 3) = 0$. Look at both terms and notice that each contains $(2y - 3)$ — that's a greatest common factor! Factor out the GCF: $(2y - 3)(y + 4) = 0$. Use the zero product property: $2y - 3 = 0$, $2y = 3$, $y = \frac{3}{2}$ and $y + 4 = 0$, $y = -4$.

2. Solve the equation $16m^2 - 8m + 1 = 0$. **The answer is $m = \frac{1}{4}$.**

How come there's only one answer? Oh right, it's a double root, probably. Let's factor it and find out: $(16m^2)(1) = 16m^2$, whose factors are m and $16m$, $-m$ and $-16m$, $2m$ and $8m$, $-2m$ and $-8m$, $4m$ and $4m$, and $-4m$ and $-4m$. The winning pair is the last one. Now, create the polynomial $16m^2 - 4m - 4m + 1 = 0$ and group it to get $4m(4m - 1) - 1(4m - 1) = 0$. Next, factor out the GCF: $(4m - 1)(4m - 1) = 0$. Notice that both factors are the same. Your answer is the same root twice! $4m - 1 = 0$, $4m = 1$, $m = \frac{1}{4}$.

3. Solve the equation $x^3 + x^2 = 9x + 9$. **The answer is $x = -3, -1,$ and 3.**

You need to get 0 on one side first: $x^3 + x^2 - 9x - 9 = 0$ will do. If you group the polynomial into two sets of two, you get two greatest common factors: $x^2(x + 1) - 9(x + 1) = 0$. This also has a GCF in it: $(x + 1)(x^2 - 9) = 0$. Notice that the right factor is a difference of squares and will factor again: $(x + 1)(x - 3)(x + 3) = 0$. Set each factor equal to 0 and solve: $x + 1 = 0$, $x = -1$; $x - 3 = 0$, $x = 3$; $x + 3 = 0$, $x = -3$.

4. Solve the equation $\frac{1}{6}x^2 + \frac{2}{3}x = 2$. **The answer is $x = 2$ and $x = -6$.**

We decided to make things different and get 0 on one side first. We're kidding, of course! You always have to get 0 on one side to solve polynomials that are second degree or higher. $\frac{1}{6}x^2 + \frac{2}{3}x - 2 = 0$. Next, we multiply every term by the least common multiple of 6 to get rid of those pesky fractions. This gives you the polynomial $x^2 + 4x - 12 = 0$. This factors to $(x + 6)(x - 2) = 0$. The zero product property gets you to the two solutions: $x = -6$ and $x = 2$.

5. Solve $x^2 - 10 = 2x$. **The answer is $1 \pm \sqrt{11}$.**

Get 0 on one side first: $x^2 - 2x - 10 = 0$. This equation doesn't factor, so use the quadratic formula to solve.

$$x = \frac{-(-2) \pm \sqrt{(-2)^2 - 4(1)(-10)}}{2(1)} = \frac{2 \pm \sqrt{4 + 40}}{2} = \frac{2 \pm \sqrt{44}}{2} = \frac{2 \pm 2\sqrt{11}}{2} = 1 \pm \sqrt{11}.$$

6. Solve $7x^2 - x + 2 = 0$. **The answer is no solution.**

This equation also doesn't factor, so use the quadratic formula to solve.

$$x = \frac{-(-1) \pm \sqrt{(-1)^2 - 4(7)(2)}}{2(7)} = \frac{1 \pm \sqrt{1 - 56}}{14} = \frac{1 \pm \sqrt{-55}}{14}.$$

That negative sign under the square root tells you that you can stop — no solution exists.

7 Solve $x^2 - 4x - 7 = 0$ by completing the square. **The answer is** $2 \pm \sqrt{11}$.

This time you're asked to complete the square. Make sure you always follow your teacher's directions. Begin by adding the 7 to both sides: $x^2 - 4x = 7$. Now factor out the leading coefficient: $1(x^2 - 4x) = 7$. Take half of –4 and square it, and add that inside the parentheses to get $1(x^2 - 4x + 4) = 7$. You need to keep the equation balanced by multiplying the coefficient and the new term you just added inside the parentheses and adding that to the opposite side. Because the coefficient is 1, that's not that hard: $1(x^2 - 4x + 4) = 7 + 4$. Now factor the trinomial: $1(x - 2)^2 = 11$, and divide the leading coefficient: $(x - 2)^2 = 11$. Square root both sides: $x - 2 = \pm \sqrt{11}$. Add the 2: $x = 2 \pm \sqrt{11}$.

8 Solve $-2.31x^2 - 4.2x + 6.7 = 0$. **The solutions are approximately –2.36 and 6.56.**

Those ugly decimals should make you reach immediately for a calculator and plug away at the quadratic formula.

$$x = \frac{-(-4.2) \pm \sqrt{(-4.2)^2 - 4(-2.31)(6.7)}}{2(-2.31)} = \frac{4.2 \pm \sqrt{17.64 + 61.908}}{-4.62} = \frac{4.2 \pm \sqrt{79.548}}{-4.62} = \frac{4.2 \pm 8.919}{2}.$$

Take your time through these types of problems. Simplify to get the final two answers: –2.36 and 6.56.

9 Solve the equation $2x^3 + 3x^2 - 18x + 8 = 0$. **The zeros are $x = -4$, ½, and 2.**

This third-degree equation has at most three real roots. The two changes in sign in f(x) show two or zero positive roots, and the one change in sign in f(–x) shows one negative root. The list of possible rational zeros is: $\pm 1, \pm \frac{1}{2}, \pm 2, \pm 4, \pm 8$.

Start off by testing $x = 2$.

$$
\begin{array}{r|rrr}
2 & 2 & 3 & -18 & 8 \\
 &   & 4 & 14 & -8 \\
\hline
 & 2 & 7 & -4 & 0 \\
\end{array}
$$

The depressed polynomial is $2x^2 + 7x - 4$, which factors to $(2x - 1)(x + 4)$, which tells you that the other two roots are $x = \frac{1}{2}$ and –4.

10 Solve the equation $12x^4 + 13x^3 - 20x^2 + 4x = 0$. **The roots are $x = 0$, ¼, –2, and ⅔.**

Factor out the GCF in all the terms first: $x(12x^3 + 13x^2 - 20x + 4) = 0$. The first factor gives you one solution immediately: $x = 0$. Now concentrate on the leftover polynomial inside the parentheses and solve: $12x^3 + 13x^2 - 20x + 4 = 0$. This has three real roots, two or zero of which are positive and one of which is negative. The list of possibilities is: $\pm 1, \pm \frac{1}{2}, \pm \frac{1}{3}, \pm \frac{1}{4}, \pm \frac{1}{6}, \pm \frac{1}{12}, \pm 2, \pm \frac{2}{3}, \pm 4$, and $\pm \frac{4}{3}$.

Start off by testing $x = -2$.

$$
\begin{array}{r|rrr}
-2 & 12 & 13 & -20 & 4 \\
 &   & -24 & 22 & -4 \\
\hline
 & 12 & -11 & 2 & 0 \\
\end{array}
$$

The depressed polynomial this time is $12x^2 - 11x + 2$. This factors to $(4x - 1)(3x - 2)$, which gets you to the last two roots: $x = \frac{1}{4}$ and $x = \frac{2}{3}$.

11 Solve the equation $x^3 + 7x^2 + 13x + 4 = 0$. **The answers are $x = -4$ and $\frac{-3 \pm \sqrt{5}}{2}$.**

This cubic has a maximum of three real roots. None of them are positive and three or one of them are negative. The list of possibilities this time (ignoring all the positives) is: –1, –2, and –4.

Start off by testing –4.

$$
\begin{array}{r|rrrr}
-4 & 1 & 7 & 13 & 4 \\
   &   & -4 & -12 & -4 \\
\hline
   & 1 & 3 & 1 & 0
\end{array}
$$

The depressed polynomial $x^2 + 3x + 1$ doesn't factor, but the quadratic formula reveals that the last two solutions are $x = \dfrac{-3 \pm \sqrt{5}}{2}$.

12 Find the roots of the equation $x^4 + 10x^3 + 38x^2 + 66x + 45$. **The roots are $x = -3$ (double root) and $x = -2 \pm i$.**

This fourth-degree polynomial also has no positive roots; 4, 2, or 0 are negative roots. The list of possibilities to pick from is: –1, –3, –5, –9, 15, –45.

Start off with $x = -3$.

$$
\begin{array}{r|rrrrr}
-3 & 1 & 10 & 38 & 66 & 45 \\
   &   & -3 & -21 & -51 & -45 \\
\hline
   & 1 & 7 & 17 & 15 & 0
\end{array}
$$

This time, when you test it again, it works.

$$
\begin{array}{r|rrrr}
-3 & 1 & 7 & 17 & 15 \\
   &   & -3 & -12 & -15 \\
\hline
   & 1 & 4 & 5 & 0
\end{array}
$$

You're left with the depressed polynomial $x^2 + 4x + 5$, which doesn't factor, but you can use the quadratic formula to find that the last two roots are imaginary: $x = -2 \pm i$.

13 If the roots of a polynomial are $x = -3, -2, 4$, and 6, what's the polynomial? **The answer is $x^4 - 5x^3 - 20x^2 + 60x + 144$.**

Use the factor theorem to help you figure this one out. If $x = -3$, then $x + 3$ is one of the factors. Similarly, if $x = -2$, then $x + 2$ is a factor; if $x = 4$, then $x - 4$ is a factor; and if $x = 6$, then $x - 6$ is a factor. If you take all the factors and multiply them, you get $(x + 3)(x + 2)(x - 4)(x - 6)$. FOIL the first two binomials to get $x^2 + 5x + 6$ and the second two binomials to get $x^2 - 10x + 24$. Multiply your way through those two polynomials:

$$
\begin{array}{r}
x^2 + 5x + 6 \\
\underline{x^2 - 10x + 24} \\
24x^2 + 120x + 144 \\
-10x^3 - 50x^2 - 60x \\
\underline{x^4 + 5x^3 + 6x^2 } \\
x^4 - 5x^3 - 20x^2 + 60x + 144
\end{array}
$$

You end up with the polynomial $x^4 - 5x^3 - 20x^2 + 60x + 144$.

14 If the roots of a polynomial are $x = 2, 4 \pm 3i$, what's the polynomial? **The answer is $x^3 - 10x^2 + 41x - 50$.**

This time the factors are $x - 2$, $x - 4 - 3i$, and $x - 4 + 3i$. In cases like these, it's easier to multiply the imaginary numbers first. When you do that you end up with the trinomial $x^2 - 8x + 25$. Now multiply that by the binomial to end up with the polynomial: $x^3 - 10x^2 + 41x - 50$.

15 Factor the polynomial $6x^4 - 7x^3 - 18x^2 + 13x + 6 = 0$. **The answer is $(x - 2)(x - 1)(3x + 1)(2x + 3)$.**

You're still using the factor theorem, but this time you have to find the roots first. The roots are $x = 2, 1, -\frac{1}{3}$, and $-\frac{3}{2}$. This means that $x - 2$, $x - 1$, $x + \frac{1}{3}$, and $x + \frac{3}{2}$ are your factors. You can get rid of those fractions by multiplying each term of the factor by the LCD. In other words, multiply $x + \frac{1}{3}$ by 3 and $x + \frac{3}{2}$ by 2. This finally gives you $(x - 2)(x - 1)(3x + 1)(2x + 3)$.

16 Factor the polynomial $x^4 + 10x^3 + 38x^2 + 66x + 45$. **The answer is $(x^2 + 4x + 5)(x + 3)^2$.**

This is problem 12. It has two imaginary roots: $x = -2 \pm i$ and $x = -3$, a double root. This means your factors are $(x + 2 + i)(x + 2 - i)(x + 3)(x + 3)$. You multiply out the two imaginary roots to come up with a polynomial factor and get $(x^2 + 4x + 5)(x + 3)^2$.

17 Graph $f(x) = x^4 + 2x^3 - 13x^2 - 14x + 24$. **See the following graph for the answer.**

The x-intercepts are: $x = -4, -2, 1$, and 3. Mark those on the graph first. Then find the y-intercept: $y = 24$. The leading coefficient test tells you that both ends of this graph point up. Here's the graph:

18 Graph $f(x) = 6x^4 - 7x^3 - 18x^2 + 13x + 6$. **See the following graph for the answer.**

You found the roots for this polynomial in problem 15: $x = -\frac{3}{2}, -\frac{1}{3}, 1$, and 2. Mark those on the graph first. Then find that $y = 6$ is the y-intercept. The leading coefficient test tells you that both ends of this graph point up. Here's the graph:

19 Graph $f(x) = 12x^4 + 13x^3 - 20x^2 + 4x$. **See the following graph for the answer.**

This is problem 10, where you found that the solutions are $x = -2$, 0, ¼, and ⅔. This polynomial has no constant, so $y = 0$ is the y-intercept. This graph crosses at the origin. The leading coefficient test tells you that both ends of the graph point up. Here's the graph:

20 Graph $f(x) = x^4 + 10x^3 + 38x^2 + 66x + 45$. **See the following graph for the answer.**

Problem 12 has roots of $x = -3$ (as a double root); the other roots are imaginary. The graph will bounce at this point. The y-intercept is 45. The leading coefficient test tells you that both ends of the graph point up. Here's the graph:

Chapter 5

Exponential and Logarithmic Functions

In This Chapter

▶ Figuring out exponential functions

▶ Looking at logarithmic functions

▶ Using exponents and logs to solve equations

▶ Working with exponential word problems

*E*xponential growth is simply the idea that something gets bigger and bigger (or smaller and smaller) very fast. Exponential and logarithmic functions can be used to describe growth or decay. They have many practical applications, such as determining population growth, measuring the acidity of a substance, and calculating financial growth. In addition, they're central to many concepts in calculus (a good reason to master them in pre-calculus!). They're different from the other functions we've been dealing with so far because their variables are no longer in the base of the expression (more on that later).

In this chapter, you practice solving equations, simplifying expressions, and graphing exponents and logarithms. In addition, you can practice manipulating functions to solve equations and practically applying the concepts to word problems.

Things Get Bigger (Or Smaller) All the Time — Solving Exponential Functions

Exponential functions are functions in which the variable is in the exponent. When the base of the exponent is greater than 1, the function gets really big really fast, and when it's less than 1, it gets really small really fast.

In exponential functions, the variable is in the power of the expression. The base can be any constant, including a special constant that mathematicians and scientists define as *e*. This irrational constant, *e*, has a value that's approximately 2.7182 and is extremely useful in exponential expressions (and in logarithms, but we're getting ahead of ourselves).

Solving exponential equations requires that you recall the basic exponent rules:

$$c^a \cdot c^b = c^{a+b}$$

$$\left(\frac{c}{d}\right)^a = \frac{c^a}{d^a}$$

$$\frac{c^a}{c^b} = c^{a-b}$$

$$c^{-a} = \frac{1}{c^a}$$

$$(c^a)^b = c^{a \cdot b}$$

$$(c \cdot d)^a = c^a \cdot d^a$$

$$c^0 = 1$$
$$\text{if } c^a = c^b \text{ then } a = b$$

While graphing exponential equations, it's important to recall the tricks for transforming graphs (see Chapter 3 for a refresher).

Q. Solve for x in $8^{4x + 12} = 16^{2x + 5}$.

A. $x = -4$. First, in order to utilize our exponential rules, it's helpful if both expressions are the same base. So, knowing that $8 = 2^3$ and $16 = 2^4$, by factoring and rewriting using exponents, you can rewrite both sides of the equation with a base of 2: $2^{3(4x + 12)} = 2^{4(2x + 5)}$. Now that your bases are the same, you can set your exponents equal to each other (using properties of exponents): $3(4x + 12) = 4(2x + 5)$. Next, you can simplify using the distributive property of equality: $12x + 36 = 8x + 20$. Finally, you can solve algebraically: $4x + 36 = 20$; $4x = -16$; $x = -4$.

Q. Sketch the graphs of (A) $y = 2^x$, (B) $y = 2^x + 1$, (C) $y = 2^{x + 3}$, (D) $y = 2^{-x}$, and (E) $y = -2^x$, all on the same set of axes.

A. Graphs B–E are all transformations of the first graph A (see Chapter 3 for a review of transformations of graphs). By adding 1 to graph A, the result is graph B, a shift up of 1 unit. By adding 3 to the exponent of graph A, the result is graph C, shifted 3 units to the left. Graph D is the result of making the exponent negative, which results in a reflection over the y-axis, and graph E, created by negating the base, results in the reflection of the graph over the x-axis. See the resulting graphs in Figure 5-1.

Figure 5-1: Transformations of the exponential equation with base 2.

1. Solve for x in $27^{x+3} = 81^{3x-9}$.

Solve It

2. Solve for x in $e^{2x-4} = e^{6x+8}$.

Solve It

3. Solve for x in $(5^{2x} - 1)(25^x - 125) = 0$.

Solve It

4. Solve for x in $3 \cdot 9^x - 8 = -7$.

Solve It

5. Sketch the graph of $y = -3^x - 4$.

Solve It

6. Sketch the graph of $y = -3e^{x-2}$.

Solve It

The Only Logs You Won't Cut: Solving Logarithms

Just as multiplying by the reciprocal is another way to write division, logarithms are simply another way to write exponents. Exponential and logarithmic functions are inverses of each other. In other words, logarithmic functions are really just another way to write exponential functions. So you may ask, "Why do you need both?" Well, logarithms are extremely helpful for an immense number of practical applications. In fact, before the invention of computers, logarithms were the only way to compute many complex computations in physics, chemistry, astronomy, and engineering.

For solving and graphing logarithmic functions (logs), remember this inverse relationship and you'll be sawing . . . er, solving logs in no time! Here's this relationship in equation form:

$$y = \log_b x \leftrightarrow b^y = x.$$

Just as with exponential functions, the base can be any number, including e. In fact, a base of e is so common in science and calculus that $\log_e$ has its own special name: ln. Thus, $\log_e x = \ln x$.

Similarly, $\log_{10}$ is so commonly used that it's just written as log (no base written).

Remember our review of domain from Chapter 3? Well, here's one of those times when domain can be tricky. The domain for the basic logarithm $y = \log_b x$ is $x > 0$. Therefore, when you're solving logarithmic functions, it's important to check for extraneous roots (review Chapter 1).

Here are more properties that are true for *any* logarithm:

$\log_b 1 = 0$

$\log_b b = 1$

The product rule: $\log_b(a \cdot c) = \log_b a + \log_b c$

The quotient rule: $\log_b\left(\dfrac{a}{c}\right) = \log_b a - \log_b c$

The power rule: $\log_b a^c = c \cdot \log_b a$

$\log_b b^x = x$

$b^{\log_b x} = x$

$\log_b a = \dfrac{\log_c a}{\log_c b}$

If $\log_b a = \log_b c$, then $a = c$

Using these properties, simplifying logarithmic expressions and solving logarithmic functions is a snap (we did say logs, not twigs, right?).

Q. Rewrite the following logarithmic expression to a single log: $3\log_5 x + \log_5(2x - 1) - 2\log_5(3x + 2)$.

A. $\log_5\left[\dfrac{x^3(2x-1)}{(3x+2)^2}\right]$

Using the properties of logs, begin by rewriting the coefficients as exponents: $3\log_5 x = \log_5 x^3$ and $2\log_5(3x + 2) = \log_5(3x + 2)^2$. Next, rewrite the addition of the first two logs as the product of logs: $\log_5 x^3 + \log_5(2x - 1) = \log_5 x^3(2x - 1)$. Last, rewrite the final log as a quotient: $\log_5 x^3(2x - 1) - \log_5(3x + 2)^2 = $

$$\log_5\left[\dfrac{x^3(2x-1)}{(3x+2)^2}\right]$$

Q. Sketch the graphs of (A) $y = \log_2 x$,
(B) $y = 1 + \log_2 x$, (C) $y = \log_2(x + 3)$, and
(D) $y = -\log_2 x$, all on the same set of axes.

A. First, in Figure 5-2, you can see that graphs
B–D are transformations of graph A. Graph
B is a shift of 1 up, graph C is a shift of 3 to
the left, and graph D is a reflection of graph
A over the *x*-axis. Second (nifty trick here),
these are all inverses of graphs A–D in the
exponential section (refer to Figure 5-1).
Another way to graph logarithms is to
change the log to an exponential using the
properties of logarithms, find the inverse
function by switching *x* and *y*, graph the
inverse, and reflect every point over the
line $y = x$. For a review of inverses, see
Chapter 3. Here, we stick with transforming
the parent graph.

Figure 5-2:
Transfor-
mations
of the loga-
rithmic
equation
with base 2.

---

7. Rewrite the given expression as a single
logarithm: $\ln 4x + 3\ln(x - 2) - 2(\ln 2x + \ln(3x - 4))$.

Solve It

8. Solve for *x* in

$$\log_7 4 + \log_7 (x + 4) - 2\log_7 2 = \log_7 (x - 2) + \frac{1}{2}\log_7 9.$$

Solve It

---

9. Solve for *x* in $\ln x + \ln(2x - 1) = \ln 1$.

Solve It

10. Find $\log_b(48b)$ if $\log_b 2 = 0.36$ and $\log_b 3 = 0.56$.

Solve It

11. Sketch the graph of $y = -3 + \log(x + 2)$.

Solve It

12. Sketch the graph of $y = \ln(x - 2) + 4$.

Solve It

Putting Them Together: Solving Equations Using Exponents and Logs

Now, we show you how to put these two lovely functions together. By keeping in mind the inverse relationship ($y = \log_b x \leftrightarrow b^y = x$), you can solve even more complex problems. Aren't you excited?!

A helpful key to remember when solving equations using exponents and logs is that if the variable is in the exponent, convert the equation into logarithmic form. This is especially helpful if you use natural log (ln) or the common log ($\log_{10} x$, often referred to as just $\log x$) because you can plug the variable into your calculator to get a decimal approximation of the solution.

One pitfall to avoid when manipulating logs relates to the products and quotients of logs. Remember: $\log 18 - \log 3 = \log\left(\dfrac{18}{3}\right)$ not $\dfrac{\log 18}{\log 3}$. These are entirely different expressions. In fact, if you plug them into your calculator, you can see that $\log\left(\dfrac{18}{3}\right) = \log 6 \approx 0.778$ while $\left(\dfrac{\log 18}{\log 3}\right) \approx 2.631$. The same can be said for products and logs: $\log 6 + \log 7 = \log(6 \cdot 7)$, not $\log(6 + 7)$.

Q. $\log(50x + 250) - \log x = 2$.

A. $x = 5$. Start by combining the logs as a quotient: $\log\left(\dfrac{50x + 250}{x}\right) = 2$. Next, rewrite in exponential form (remember that log

means $\log_{10}$): $\dfrac{50x + 250}{x} = 10^2$. Because $10^2 = 100$, you can rewrite the equation as $\dfrac{50x + 250}{x} = 100$. After cross-multiplying, you can then solve algebraically: $50x + 250 = 100x$; $250 = 50x$; $x = 5$.

EXAMPLE

Q. Solve for x in $3^x = 2^{(x+2)}$.

A. $x = \dfrac{\ln 4}{\ln\left(\dfrac{3}{2}\right)} \approx 3.419$. First, recognize that the variable is in the exponent of each term, so you can easily remedy that by taking either log or ln of both sides. We're going to use ln, but it really doesn't make a difference. So $3^x = 2^{(x+2)}$ becomes $\ln 3^x = \ln 2^{(x+2)}$. Then, you can use properties of logarithms to solve. Start by changing the exponents to coefficients: $x \cdot \ln 3 = (x+2)\ln 2$. Using algebra, you can distribute the ln2 across $(x+2)$: $x \cdot \ln 3 = x \cdot \ln 2 + 2 \cdot \ln 2$. Still using algebra, get the terms with the variable on the same side by subtracting $x \cdot \ln 2$ to the opposite side: $x \cdot \ln 3 - x \cdot \ln 2 = 2 \cdot \ln 2$. Then, using distributive property again, remove the x as a greatest common factor: $x(\ln 3 - \ln 2) = 2 \cdot \ln 2$. Finally, isolate x by dividing $\ln 3 - \ln 2$ from both sides: $x = \dfrac{\ln 2^2}{\ln 3 - \ln 2}$. Last, use the quotient rule to simplify and get $x = \dfrac{\ln 4}{\ln\left(\dfrac{3}{2}\right)} \approx 3.419$.

---

13. Solve for x in $\log(x-6) - \log(x+3) = 1$.

Solve It

14. Solve for x: $3^x = 5$.

Solve It

---

15. Solve for x in $4^x - 4 \cdot 2^x = -3$.

Solve It

16. Solve for x: $3^x = 5^{(2x-3)}$.

Solve It

Solving Exponential Word Problems . . . Argh!

When will I ever use this? Well, in addition to using exponential functions in a mathematics course, they actually have many practical applications. Common uses of exponential functions include figuring compound interest, computing population growth, and doing radiocarbon dating (no, not some new online matchmaking system). In fact, these are so common, many teachers make you memorize their formulas. If you need nonstandard formulas to do a problem, they'll be provided in the question itself.

Here are formulas for interest rate and half-life:

- ✔ Compound interest formula: $A = P\left(1+\dfrac{r}{n}\right)^{nt}$ where A is the amount after t time compounded n times per year if P dollars are invested at interest rate r.

- ✔ Continuous compound interest formula: $A = Pe^{r \cdot t}$, where A is the amount after t time if P dollars are invested at interest rate r continuously throughout the year.

- ✔ Formula for the half-life of a radioactive element: $M(x) = c \cdot 2^{\frac{-x}{h}}$ where $M(x)$ is the mass at the time x, c is the original mass of the element, and h is the half-life of the element.

Q. If you deposit $600 at 5.5% interest compounded continuously, what will your balance be in 10 years?

A. **$1,039.95.** Because this is continuous compound interest, you use the formula $A = Pe^{rt}$ when you're solving for A. $A = \$600e^{(0.055)(10)}$. Plugging this into a calculator, you get approximately $1,039.95.

Q. How old is a piece of bone that has lost 60% of its carbon-14 (half-life of carbon-14 is 5,730 years)?

A. **Approximately 7,575 years old.** We can figure out this problem using the formula for half-life. First, because 60% of the carbon-14 is gone, the mass of carbon remaining is 40%, so we can write the present mass as $.40c$. Therefore, the equation will be: $0.40c = c \cdot 2^{\frac{-x}{5,730}}$. We can start solving this by cancelling c from both sides: $0.40 = 2^{\frac{-x}{5,730}}$. Taking the natural log of both sides allows us to move the variable from the exponent position: $\ln 0.40 = \ln 2^{\frac{-x}{5,730}}$,

then $\ln 0.40 = \left(\dfrac{-x}{5,730}\right)\ln 2$. From here, we

can solve algebraically: $\dfrac{\ln 0.40}{\ln 2} = \dfrac{x}{-5,730}$;

$-5,730\dfrac{\ln 0.40}{\ln 2} \approx 7,575\,years$.

17. If you deposit $3,000 at 8% interest per year compounded quarterly, in approximately how many years will the investment be worth $10,500?

Solve It

18. The half-life of Krypton-85 is 10.4 years. How long will it take for 600 grams to decay to 15 grams?

Solve It

19. The deer population in a certain area in year t is approximately $P(t) = \dfrac{3,000}{1 + 299 \cdot e^{-0.56t}}$. When will the deer population reach 2,000?

Solve It

20. If you deposit $20,000 at 6.5% interest compounded continuously, how long will it take for you to have $1,000,000?

Solve It

Answers to Problems on Exponential and Logarithmic Functions

1 Solve for x in $27^{x+3} = 81^{3x-9}$. **The answer is $x = 5$.**

First, rewrite 27 as 3^3 and 81 as 3^4. Simplify the power to get $3^{3(x+3)} = 3^{4(3x-9)}$. Now that the bases are the same, set the two exponents equal to each other: $3(x+3) = 4(3x-9)$, and then solve for x: $3x + 9 = 12x - 36$; $-9x = -45$; $x = 5$.

2 Solve for x in $e^{2x-4} = e^{6x+8}$. **The answer is $x = -3$.**

Start by setting the exponents equal to each other: $2x - 4 = 6x + 8$; then solve algebraically: $4x = -12$; $x = -3$.

3 Solve for x: $(5^{2x} - 1)(25^x - 125) = 0$. **The answer is $x = 0, \frac{3}{2}$.**

Using the fact that $25 = 5^2$, replace 25^x with 5^{2x} to get $(5^{2x} - 1)(5^{2x} - 125) = 0$. Next, set each factor equal to 0 using the zero product property (see Chapter 4 for a review) and solve: First, $5^{2x} - 1 = 0$, $5^{2x} = 1$, and, because anything to the power of 0 equals 1, $5^{2x} = 5^0$. Therefore, $2x = 0$, $x = 0$. Second, $5^{2x} - 125 = 0$, $5^{2x} = 125$, and because 125 is equal to 5^3, rewrite the second equation as $5^{2x} = 5^3$. Set the exponents equal to each other: $2x = 3$, and solve for $x = \frac{3}{2}$. Both solutions work.

4 Solve for x in $3 \cdot 9^x - 8 = -7$. **The answer is $x = -\frac{1}{2}$.**

Start by isolating the exponential expression: $3 \cdot 9^x = 1$; $9^x = \frac{1}{3}$. Next, replace 9^x with 3^{2x} and $\frac{1}{3}$ with 3^{-1}, so $3^{2x} = 3^{-1}$. Set the exponents equal to each other: $2x = -1$, and solve for $x = -\frac{1}{2}$.

5 Sketch the graph of $y = -3^x - 4$.

The y-intercept is $(0, -5)$. The graph of this function is the basic exponential graph of $y = 3^x$ shifted 4 units down and reflected upside down.

6 Sketch the graph of $y = -3e^{x-2}$.

The y-intercept is $(0, -0.406)$. The graph of this function is the basic exponential graph of $y = e^x$ shifted 2 units to the right, reflected upside down, and contracted by a unit of 3.

7 Rewrite the given expression as a single logarithm: $\ln 4x + 3\ln(x-2) - 2(\ln 2x + \ln(3x-4))$. **The answer is** $\ln\left[\dfrac{(x-2)^3}{x(3x-4)^2}\right]$.

Begin by rewriting coefficients as exponents: $\ln 4x + \ln(x-2)^3 - (\ln(2x)^2 + \ln(3x-4)^2)$. Next, rewrite the first two logarithms as a single product: $\ln 4x(x-2)^3 - (\ln 4x^2 + \ln(3x-4)^2)$. Then use the product rule in the parentheses to get $\ln 4x(x-2)^3 - (\ln 4x^2(3x-4)^2$. Finally, write the difference of logarithms as a quotient and reduce the 4 in the numerator and denominator:

$$\ln\left[\dfrac{(x-2)^3}{x(3x-4)^2}\right].$$

8 Solve for x in: $\log_7 4 + \log_7(x+4) - 2\log_7 2 = \log_7(x-2) + \tfrac{1}{2}\log_7 9$. **The answer is $x = 5$.**

The first step is to write the coefficients as exponents:

$\log_7 4 + \log_7\left(x+4\right) - \log_7 2^2 = \log_7\left(x-2\right) + \log_7 9^{\frac{1}{2}}$. Next, rewrite the sums and differences as the

products and quotients of logarithms: $\log_7 \dfrac{4\left(x+4\right)}{2^2} = \log_7\left(x-2\right)\cdot 9^{\frac{1}{2}}$. Using the rules of logarithms,

set $\dfrac{4\left(x+4\right)}{2^2} = \left(x-2\right)\cdot 9^{\frac{1}{2}}$. Solve algebraically: $\dfrac{4\left(x+4\right)}{2^2} = \left(x-2\right)\cdot 9^{\frac{1}{2}}; \dfrac{4\left(x+4\right)}{4} = \left(x-2\right)\cdot\sqrt{9}; x+4 =$

$3(x-2); x+4 = 3x-6; 2x = 10; x = 5$.

9 Solve for x in: $\ln x + \ln(2x-1) = \ln 1$. **The answer is $x = 1$.** Rewriting the sum of natural logs as a product, we get: $\ln x(2x-1) = \ln 1$. Then, using rules of logarithms, set $x(2x-1) = 1$ and solve algebraically: $2x^2 - x = 1, 2x^2 - x - 1 = 0$. Factor the quadratic: $(2x+1)(x-1) = 0$, and using the zero product property, set each factor equal to 0. The solutions are $x = -\tfrac{1}{2}$ and $x = 1$, but $-\tfrac{1}{2} < 0$ so it's extraneous, and the only solution is $x = 1$.

10 Find $\log_b(48b)$ if $\log_b 2 = 0.36$ and $\log_b 3 = 0.56$. **The answer is 3.**

Start by expanding the logarithm into the sum of two logs: $\log_b 48 + \log_b b$. Next, factoring the 48 into $16 \cdot 3$, expand again: $\log_b 16 + \log_b 3 + \log_b b$. Using the fact that $16 = 2^4$, write the first log: $\log_b 2^4 + \log_b 3 + \log_b b$. Then, write the exponent as a coefficient: $4\log_b 2 + \log_b 3 + \log_b b$. Last, replace $\log_b 2$ with 0.36, $\log_b 3$ with 0.56, and $\log_b b$ with 1 and simplify: $4(0.36) + 0.56 + 1 = 3$.

11 Sketch the graph of $y = -3 + \log(x + 2)$.

The y-intercept is $(0, -2.699)$. The graph of this function is the basic exponential graph of $y = \log x$ shifted 2 units to the left and 3 units down.

12 Sketch the graph of $y = \ln(x - 2) + 4$.

There's a vertical asymptote at $x = 2$. The graph of this function is the basic exponential graph of $y = \ln x$ shifted 2 units to the right and 4 units up.

13 Solve for x in $\log(x + 6) - \log(x - 3) = 1$. **The answer is $x = 4$.**

Begin by writing the difference of logs as a quotient: $\log\dfrac{x+6}{x-3} = 1$. Next, rewrite the logarithm as

an exponent: $\dfrac{x+6}{x-3} = 10^1$. Then, solve algebraically: $x + 6 = 10(x - 3)$; $x + 6 = 10x - 30$; $36 = 9x$; $x = 4$.

14 Solve for x: $3^x = 5$. **The answer is $x = \dfrac{\ln 5}{\ln 3}$ or $x \approx 1.464$.**

First, take the natural log of both sides: $\ln 3^x = \ln 5$. Then use the power rule to simplify: $x \cdot \ln 3 = \ln 5$.

Last, divide both sides by $\ln 3$: $x = \dfrac{\ln 5}{\ln 3}$ or $x \approx 1.464$.

15 Solve for x in: $4^x - 4 \cdot 2^x = -3$. **The answer is $x = \dfrac{\ln 3}{\ln 2} \approx 1.585$ and $x = 0$.**

Start by using the fact that $4 = 2^2$ and rewrite 4^x as 2^{2x}: $2^{2x} - 4 \cdot 2^x = -3$. Add 3 to both sides: $2^{2x} - 4 \cdot 2^x + 3 = 0$. Notice that this is the same thing as $(2^x)^2 - 4 \cdot 2^x + 3$ so that you can substitute y for 2^x: $y^2 - 4y + 3 = 0$. Now, you can factor and solve using zero product property: $(y-3)(y-1) = 0$; $y = 3$ and $y = 0$. Then, resubstitute 2^x for y: $2^x = 3$ and $2^x = 1$. Taking the natural log of each side, you can solve for x by using the rules of logarithms: $\ln 2^x = \ln 3$; $x\ln 2 = \ln 3$; $x = \dfrac{\ln 3}{\ln 2}$, your first solution.

Finally, $\ln 2^x = \ln 1$; $x\ln 2 = \ln 1$; $x = \dfrac{\ln 1}{\ln 2}$; because $\ln 1 = 0$, $\dfrac{\ln 1}{\ln 2} = 0$, your second solution. Both solutions work.

16 Solve for x: $3^x = 5^{(2x-3)}$: **The answer is $x = \dfrac{\ln \frac{1}{125}}{\ln \frac{3}{25}} \approx 2.28$.**

First, take the natural log of both sides: $\ln 3^x = \ln 5^{(2x-3)}$. Then use properties of logarithms to solve. Start by changing the exponents to coefficients: $x \cdot \ln 3 = (2x-3)\ln 5$. Using algebra, you can distribute the $\ln 5$ across $(2x-3)$: $x \cdot \ln 3 = 2x \cdot \ln 5 - 3 \cdot \ln 5$. Still using algebra, get the terms with the variable on the same side by subtracting $2x \cdot \ln 5$ to the opposite side: $x \cdot \ln 3 - 2x \cdot \ln 5 = -3 \cdot \ln 5$. Then, using distributive property again, remove the x as a greatest common factor:

$x(\ln 3 - 2\ln 5) = -3 \cdot \ln 5$. Combine the difference of logarithms as a quotient: $x \cdot \ln\left(\dfrac{3}{5^2}\right) = -3\ln 5$.

Next, isolate x using algebra: $x = \dfrac{-3 \cdot \ln 5}{\ln \frac{3}{25}}$. Finally, simplify your answer using the rules of expo-

nents: $\dfrac{\ln 5^{-3}}{\ln \frac{3}{25}} = \dfrac{\ln \frac{1}{125}}{\ln \frac{3}{25}}$, which equals approximately 2.28.

17 If you deposit \$3,000 at 8% interest per year, compounded quarterly, in approximately how many years will the investment be worth \$10,500? **The answer is approximately 15.82 years.**

Using the equation: $A = P\left(1 + \dfrac{r}{n}\right)^{nt}$, where $A = \$10,500$, $P = \$3,000$, r (as a decimal) $= 0.08$, and $n = 4$:

$10,500 = 3,000\left(1 + \dfrac{0.08}{4}\right)^{4t}$; $10,500 = 3,000\,(1.02)^{4t}$; $3.5 = (1.02)^{4t}$; using logarithms: $\log 3.5 = \log 1.02^{4t}$;

$\log 3.5 = 4t\log 1.02$; finally, solve algebraically: $4t = \dfrac{\log 3.5}{\log 1.02}$; $t = t = \dfrac{1}{4} \cdot \dfrac{\log 3.5}{\log 1.02}$, which equals approximately 15.82 years.

18 The half-life of Krypton-85 is 10.4 years. How long will it take for 600 grams to decay to 15 grams? **The answer is 55.3 years.**

Using the half-life formula: $M(x) = c \cdot 2^{\frac{-x}{h}}$, where $M(x) = 15$ grams, the original mass c is 600

grams, and the half-life h is 10.4, we get $15 = 600 \cdot 2^{\frac{-x}{10.4}}$. Simplifying to $0.025 = 2^{\frac{-x}{10.4}}$, we can solve

using logarithms: $\ln 0.025 = \ln 2^{\frac{-x}{10.4}}$, $\ln 0.025 = \left(\dfrac{-x}{10.4}\right)\ln 2$, $\dfrac{\ln 0.025}{\ln 2} = \dfrac{x}{-10.4}$, and $-10.4 \cdot \dfrac{\ln 0.025}{\ln 2} = x$, which approximates to 55.3 years.

19 The deer population in a certain area in year t is approximately $P(t) = \dfrac{3,000}{1 + 299 \cdot e^{-0.56t}}$. When will the deer population reach 2,000? **The answer is approximately 11.4 years.**

Here, you simply plug in 2,000 for $P(t)$ and solve: $2,000 = \dfrac{3,000}{1 + 299 \cdot e^{-0.56t}}$; $2,000(1 + 299e^{-0.56t}) = 3,000$;

$1 + 299^{-0.56t} = 1.5$; $299e^{-0.56t} = 0.5$; $e^{-0.56t} = 0.00167$; $-0.56t = \ln 0.00167$; $t = \dfrac{\ln 0.00167}{-0.56}$, which equals approximately 11.4 years.

20 If you deposit $20,000 at 6.5% interest compounded continuously, how long will it take for you to have $1,000,000? **The answer is approximately 24.8 years.**

Using the equation for continuous compound interest: $A = Pe^{r \cdot t}$, where the amount A is $1,000,000, the initial investment P is $20,000, and the interest rate r in decimal form is 0.065, we get: $1,000,000 = 20,000e^{0.065t}$. Simplifying and using logarithms to solve: $50 = e^{0.065t}$; $\ln 50 = 0.065t$; $\frac{\ln 5}{0.065} = t$, which equals approximately 60.2 years.

Part II

Trig Is the Key: Basic Review, the Unit Circle, and Graphs

The 5th Wave   By Rich Tennant

"Periodic functions?! I can tell you all about periodic functions. My whole life is about periodic functions!"

In this part . . .

You should be familiar with the basics of trigonometry from your Geometry class — right triangles, trig ratios, and angles, for example. But your Algebra II course may or may not have expanded on those ideas to prepare you for the direction that pre-calc is going to take you. For this reason, we assume that you've never seen this stuff before. We don't want to leave you behind on our journey.

This part begins with trig ratios and word problems and then moves on to the unit circle: how to build it and how to use it. We solve some trig equations and make and measure arcs. Graphing trig functions is a major component of pre-calc, so we walk you through how to graph each of the six functions.

Chapter 6

Basic Trigonometry and the Unit Circle

In This Chapter

▶ Working with the six trigonometric ratios

▶ Making use of right triangles to solve word problems

▶ Using the unit circle to find points, angles, and right triangle ratios

▶ Isolating trig terms to solve trig equations

▶ Calculating arc lengths

Ah . . . trigonometry, the math of triangles! Invented by the ancient Greeks, trigonometry is used to solve problems in navigation, astronomy, and surveying. Think of a sailor lost at sea. All he has to do is triangulate his position against two other objects, such as two stars, and calculate his position using — you guessed it — trigonometry!

In this chapter, we review the basics of right triangle trigonometry. Then we show you how to apply it to the unit circle, a very useful tool for graphically representing trigonometric ratios and relationships. From there, you can solve trig equations. Finally, we combine these concepts so that you can apply them to arcs. The ancient Greeks didn't know what they started with trigonometry, but the modern applications are endless!

It's All Right-Triangle Trig — Finding the Six Trigonometric Ratios

Dude! Did you see that? He just did a 2π on his board! Huh? Oh . . . we mean a 360. In geometry, angles are measured in degrees (°), with 360° describing a full circle on a coordinate plane (or skateboard). However, in pre-calculus, you also use another measure for angles: radians. Radians, from the word radius, are usually designated without a symbol. Because both radians and degrees are used often in pre-calc, you see both used here.

To convert radians to degrees and vice versa, you use the fact that 360° = 2π radians, or 180° = π radians. Therefore, to convert degrees to radians, you simply multiply by the ratio $\pi/_{180°}$. Similarly, to convert radians to degrees, you can multiply by the ratio $^{180°}/_\pi$.

When solving right triangles or finding all the sides and angles (θ), it's important to remember the six basic trigonometric ratios: sine (sinθ), cosine (cosθ), tangent (tanθ), cosecant (cscθ), secant (secθ), and cotangent (cotθ). The first three are the most important to remember, as the second three are inverses of the first. In other words, sinθ = $^1/_{cscθ}$, cosθ = $^1/_{secθ}$, and tanθ = $^1/_{cotθ}$.

TIP

Acronyms! We love 'em . . . LOL — otherwise known as Laugh Out Loud! One of the most famous acronyms in math is SOHCAHTOA. It helps you remember the first three trigonometric ratios:

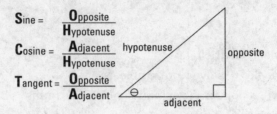

$$\mathbf{S}\text{ine} = \frac{\mathbf{O}\text{pposite}}{\mathbf{H}\text{ypotenuse}}$$

$$\mathbf{C}\text{osine} = \frac{\mathbf{A}\text{djacent}}{\mathbf{H}\text{ypotenuse}}$$

$$\mathbf{T}\text{angent} = \frac{\mathbf{O}\text{pposite}}{\mathbf{A}\text{djacent}}$$

EXAMPLE

Q. Given △KLM in Figure 6-1, find sin∠K.

Figure 6-1: Finding sin∠K in △KLM.

A. **sin∠K = $\frac{5\sqrt{34}}{34}$**. Because sin∠K = $^{\text{opp}}/_{\text{hyp}}$,

you first need to find the hypotenuse. To do this, you need to use the Pythagorean Theorem, which says that $(\text{leg})^2 + (\text{leg})^2 = (\text{hypotenuse})^2$. Using this, you find that $3^2 + 5^2 = (\text{hyp})^2$, and therefore $34 = (\text{hyp})^2$, so the hypotenuse is $\sqrt{34}$. Plugging this into your sine ratio, you get sin∠K = $\frac{5}{\sqrt{34}}$.

But if you recall what we review in Chapter 2, you need to rationalize this fraction. Doing so, you get sin∠K = $\frac{5\sqrt{34}}{34}$.

Q. Solve △RST (see Figure 6-2).

Figure 6-2: Solving △RST.

A. **RT = $\sqrt{13}$, ∠T = 33.7°, and ∠R = 56.3°.**

Remember, solving a triangle means finding all the angles and sides. So you start by using the Pythagorean Theorem to find the hypotenuse: $2^2 + 3^2 = (\text{hyp})^2$, hyp = $\sqrt{13}$.

Next, use any trigonometric ratio to find an angle. You can use sin∠T = $\frac{2}{\sqrt{13}}$. To get the angle by itself, you use the fact that the inverse operation of sin is $\sin^{-1}$, or arcsine.

Thus, you get ∠T = $\sin^{-1}\left(\frac{2}{\sqrt{13}}\right)$, which

you can find using your calculator to be 33.7°. Or, if you want to use radians, ∠T is 0.59. We prefer degrees for now. Lastly, using the fact that the angles of a triangle add up to 180°, you can find ∠R: $180 - (90 + 33.7°) = 56.3°$.

1. Find cosA in △ABC.

Solve It

2. Solve △DEF.

Solve It

3. Find ∠Q in △QRS (round to the nearest tenth).

Solve It

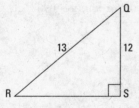

4. Find the six trigonometric ratios of ∠R in △QRS from question 3.

Solve It

Solving Word Problems with Right Triangles

Uh-oh! The dreaded word problems! Don't let these scare you. We've got some easy steps to help you through them. First, as with most word problems in math, we suggest you draw a picture. That way you can visualize the problem and it's not as scary. Second, remember that these are just right triangles. Therefore, all you have to do is use what you already know about right triangles to solve the problems. Simple!

Angle of elevation and *angle of depression* (see Figure 6-3) are two terms that come up often in right triangle word problems. They just refer to whether the angle rises from the horizon, *angle of elevation*, or falls from the horizon, *angle of depression*.

Figure 6-3:
The ups and downs: angle of elevation and angle of depression.

Q. When the sun is at an angle of elevation of 32°, a building casts a shadow that's 152 feet from its base. What is the height of the building?

A. **The building is approximately 95 feet tall.** Okay, remember your steps. Step one, draw a picture:

Step two, recall what you know about right triangles. Because you want to find the building's height, which is opposite the angle, and you have the shadow length, which is adjacent to the angle, you can use the tangent ratio. Setting up your ratio, you get $\tan 32° = \frac{x}{152}$, or $x = 152 \cdot \tan 32$. Using a calculator, you find that the building height is approximately 95 feet.

Q. Two boat captains whose boats are in a straight line from a lighthouse look up to the top of the lighthouse at the same time. The captain of boat A sees the top of the 40-foot lighthouse from an angle of elevation of 45°, while the captain of boat B sees the top of the lighthouse from an angle of elevation of 30°. How far are the boats from each other, to the nearest foot?

A. **The boats are 29 feet apart.** Ooh . . . this is a tricky one! But don't let it scare you — it's completely doable! First, remember to draw a picture. In this case, you may want to draw three: one for the lighthouse and both boats, and two separate pictures each with one boat:

From the picture, you can see that to find the distance between the boats, you need to find the distance that each boat is from the base of the lighthouse and subtract boat A's distance from the distance of boat B. Because the angle of elevation is 45° for boat A, you can set up the trigonometric ratio: $\tan 45° = \frac{40}{a}$. Solving for a, you find that the distance from boat A to the base of the lighthouse is 40 feet. Similarly, you can set up a trigonometric ratio for boat B's distance: $\tan 30° = \frac{40}{b}$. Solving, you get that $b = 69$ feet. Subtracting these two distances, you find that the distance between the boats is 29 feet. Whew!

5. Romero wants to deliver a rose to his girlfriend, Jules, who is sitting on her balcony 24 feet above the street. If Romero has a 28-foot ladder, at what angle must he place the bottom of the ladder to reach his love, Jules?

Solve It

6. Sam needs to cross a river. He spies a bridge directly ahead of him. Looking across the river, he sees that he's 27° below the bridge from the other side. How far must he walk on his side of the river to reach the bridge if the bridge length is 40 feet?

Solve It

7. Paul, a 6-foot-tall man, is holding the end of a kite string 5 feet above the ground. The string of the kite is 75 feet long at 35° of elevation. Paulette, Paul's 5-foot-tall daughter, is directly underneath the kite. How far above Paulette's head is the kite?

Solve It

8. To hold up a 100-foot pole, two guide wires are put up on opposite sides of the pole. One wire makes a 36° angle with the ground and the other makes a 47° angle with the ground. How far apart are the bases of the wires?

Solve It

Unit Circle and the Coordinate Plane: Finding Points and Angles

The unit circle is a very useful tool in pre-calculus. The information it provides can help you solve problems very quickly. Essentially, the *unit circle* is a circle with a radius (r) of one unit, centered on the origin of a coordinate plane. If you think of the trigonometric ratios you've been dealing with in terms of x and y values, where x is adjacent to the angle, y is opposite the angle, and r is the hypotenuse, that allows you to make a right triangle by using a point on the unit circle and the x-axis. This is often called *point-in-plane*, and it results in an alternate definition of the six trigonometric ratios:

⤴ $\sin\theta = y/r$

⤴ $\cos\theta = x/r$

⤴ $\tan\theta = y/x$

⤴ $\csc\theta = r/y$

⤴ $\sec\theta = r/x$

⤴ $\cot\theta = x/y$

When graphing on a coordinate plane, it's important how you measure your angles. In pre-calculus, the angle always begins on the positive side of the x-axis, called the *initial side*. Any angle in this position is in *standard position*. The angle can extend to anywhere on the plane, ending on what's called the *terminal side*. Any angles that have different measures but have the same terminal side are called *co-terminal angles*. These can be found by adding or subtracting 360° or 2π to any angle.

From the initial side, an angle that moves in the counterclockwise direction has a *positive measure*, and an angle that moves in the clockwise direction has a *negative measure*.

Q. Find three co-terminal angles of 520°.

A. **Sample answers: 160°, –200°, and 880°, but other answers are possible.** To get these, you simply add or subtract multiples of 360° from 520°. 520° – 360° = 160°; 520° – 2 · 360° = –200°; and 520° + 360° = 880°.

Q. Evaluate the six trigonometric ratios of the point (2, –3).

A. $\sin\theta = \dfrac{-3\sqrt{13}}{13}$, $\cos\theta = \dfrac{2\sqrt{3}}{13}$, $\tan\theta = {}^{-3}\!/_2$,

$\csc\theta = \dfrac{\sqrt{13}}{-3}$, $\sec\theta = \dfrac{\sqrt{13}}{2}$, $\cot\theta = {}^2\!/_{-3}$.

Start by finding the radius using the Pythagorean Theorem: $2^2 + (-3)^2 = r^2$, $4 + 9 = r^2$, $\sqrt{13} = r^2$, $\sqrt{13} = r$. Then, simply plug into the trigonometric ratios given $x = 2$, $y = -3$, and $r = \sqrt{13}$. Don't forget to rationalize any radicals in the denominator!

$\sin\theta = \dfrac{-3}{\sqrt{13}} = \dfrac{-3\sqrt{13}}{13}$, $\cos\theta = \dfrac{2}{\sqrt{13}} = \dfrac{2\sqrt{13}}{13}$,

$\tan\theta = {}^{-3}\!/_2$, $\csc\theta = \dfrac{\sqrt{13}}{-3}$, $\sec\theta = \dfrac{\sqrt{13}}{2}$,

$\cot\theta = {}^2\!/_{-3}$.

9. Find three co-terminal angles of $\dfrac{\pi}{5}$.

Solve It

10. Find two positive co-terminal angles of –775°.

Solve It

11. Evaluate the six trigonometric ratios of the point (3, 4).

Solve It

12. Evaluate the six trigonometric ratios of the point (–5, –7).

Solve It

13. Evaluate the six trigonometric ratios of the point $(-2, \ 2\sqrt{3})$.

Solve It

14. Evaluate the six trigonometric ratios of the point $(6, \ -3\sqrt{5})$.

Solve It

Isn't That Special? Finding Right Triangle Ratios on the Unit Circle

Well, isn't that special? Yes, it *is* special — special right triangles that is! Remember your geometry teacher drilling in 30-60-90 and 45-45-90 triangles? Well, they're back! And with good reason, as they're very common in pre-calculus, and they're the foundation of the unit circle. Using these two special triangles, you can find the specific trig values that you can see on a completed unit circle (to see one, check out the Cheat Sheet in the front of this book).

One important point to remember about the unit circle is that the radius is 1. Therefore, the hypotenuse of any right triangle drawn from a point to the *x*-axis is 1. Thus, for any point, (x, y), you know that $x^2 + y^2 = 1$.

Recalling 30-60-90 triangles, the sides are in the ratio of $1 : \sqrt{3} : 2$. Therefore, if you want the hypotenuse to be 1, as it is in the unit circle, divide each side by 2. Similarly, the sides of 45-45-90 triangles are in the ratio of $1 : 1 : \sqrt{2}$. Converting to a unit circle, the values are $\frac{\sqrt{2}}{2} : \frac{\sqrt{2}}{2} : 1$.

Now, using the point-in-plane definition, the six trigonometric ratios are easy to find. In fact, because the hypotenuse is now 1, $\sin\theta = \frac{y}{r}$ now becomes $\sin\theta = x$. Similarly, $\cos\theta = \frac{x}{r}$ now becomes $\cos\theta = y$. Thus, any point on the unit circle is now $(\cos\theta, \sin\theta)$. Imagine the possibilities!

If you don't have a unit circle handy, you can always use *reference angles* (θ') to find your solutions. A reference angle is the angle between the *x*-axis and the terminal side of an angle. It's different for each quadrant (see Figure 6-4). If the original angle is θ, then the reference angle in quadrant I is θ. In quadrant II, the reference angle is $180° - \theta$. For quadrant III, the reference angle is $\theta - 180°$. Lastly, quadrant IV's reference angle is $360° - \theta$.

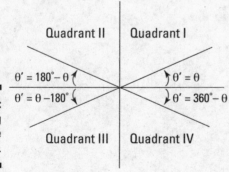

Figure 6-4: Referencing reference angles.

Q. Evaluate the six trigonometric ratios of 120° on the unit circle.

A. $\sin\theta = \frac{\sqrt{3}}{2}$, $\cos\theta = -\frac{1}{2}$, $\tan\theta = -\sqrt{3}$,

$\csc\theta = \frac{2\sqrt{3}}{3}$, $\sec\theta = -2$, and $\cot\theta = -\frac{\sqrt{3}}{3}$.

Start by finding the reference angle on th3e unit circle. Knowing the angle is in quadrant II, the reference angle is 180° − 120°, or 60°. You now have a 30-60-90 triangle! Therefore, $x = -\frac{1}{2}$ and $y = \frac{\sqrt{3}}{2}$. Now you can easily find the trig ratios using the point-in-plane definition. Keep in mind that $r = 1$. You start with sine, cosine, and tangent: $\sin\theta = y = \frac{\sqrt{3}}{2}$ (look at the definition of y earlier in this explanation — that's correct, as is the sin at the beginning of the answer); $\cos\theta = x = -\frac{1}{2}$; $\tan\theta = \frac{y}{x} = -\frac{1}{2} \div \frac{\sqrt{3}}{2} = -\frac{1}{2} \cdot \frac{2}{\sqrt{3}} = -\frac{1}{\sqrt{3}} = -\frac{\sqrt{3}}{3}$. Next, do the reciprocal ratios: $\csc\theta = \frac{1}{y} = \frac{2}{\sqrt{3}} = \frac{2\sqrt{3}}{3}$; $\sec\theta = \frac{1}{x} = 1 \div \frac{1}{2} = 1 \cdot \frac{2}{1} = -2$; and $\cot\theta = \frac{x}{y} = \frac{\sqrt{3}}{2} \div +\frac{1}{2} = \frac{\sqrt{3}}{2} \cdot \frac{2}{1} = -\sqrt{3}$.

Q. What's the value of θ when sinθ = ½ and 90° < θ < 360°?

A. **θ = 150°.** On a unit circle, sinθ = ½ when θ = 30° and 150°. Because you're limited to 90° < θ < 360°, the answer is just 150°.

15. Evaluate the six trigonometric ratios of 225°.

Solve It

16. Find θ when $\cos\theta = \frac{\sqrt{3}}{2}$ and 0° < θ < 360°.

Solve It

17. Evaluate the six trigonometric ratios of 330°.

Solve It

18. Find θ when $\tan\theta = \dfrac{\sqrt{3}}{3}$ and $0° < \theta < 360°$.

Solve It

Solving Trig Equations

Solving trigonometric equations is just like solving regular algebraic equations, with a twist! The twist is the trigonometric term. Now, instead of isolating the variable, you need to isolate the trigonometric term. From there, you can use the handy unit circle to find your solution. For a complete unit circle, check out the Cheat Sheet in the front of this book.

Given what you already know about co-terminal angles, you know that any given equation may have an infinite number of solutions. Therefore, for these examples, you stick with angles that are within one positive rotation of the unit circle $0 \le x \le 2\pi$. But be sure that you check for multiple solutions within that unit circle!

EXAMPLE

Q. Solve $2\sin x = 1$ in terms of radians.

A. $x = \frac{\pi}{6}$, **or** $\frac{5\pi}{6}$. Because you already know how to solve $2y = 1$, you also know how to solve $2\sin x = 1$ — it's $\sin x = \frac{1}{2}$. The question is, what do you do with it from there? Well, now you need to find the angle or angles that make the equation true. Here's where that unit circle comes in handy! Remembering that $\sin \theta = y$, you can look at the unit circle to find which angles have $y = \frac{1}{2}$. The two angles are $\frac{\pi}{6}$ and $\frac{5\pi}{6}$.

Q. Solve $2\cos^2 x - \cos x = 1$, giving answers in terms of degrees.

A. $x = 0°, 120°, 240°,$ **and** $360°$. Don't let this one trip you up! Just keep in mind your amazing basic algebra skills. If you think of this as $2y^2 - y = 1$, you see that it's a simple quadratic that you need to try and factor, and then use the zero product property to solve: $2\cos 2x - \cos x - 1 = 0$ factors into $(2\cos x + 1)(\cos x - 1) = 0$. Using the zero product property: $2\cos x + 1 = 0$, so $2\cos x = -1$ and $\cos x = -\frac{1}{2}$, or $\cos x - 1 = 0$, so $\cos x = 1$. Now it's time to use those reference angles! Ask yourself, when is $\cos x = -\frac{1}{2}$? Well, considering that $\cos \theta = x$ on the unit circle, it's clear that your reference angle (θ') is $60°$, and your answer falls in quadrants II and III. Therefore, the resulting angles are in quad II $(180° - 60° = 120°)$ and quad III $(180° + 60° = 240°)$. For your second equation, $\cos x = 1$ and $x = 0°$ or $360°$, because the only place on the unit circle that $x = 1$ is on the x-axis. Therefore, your four solutions are $0°, 120°, 240°,$ and $360°$.

19. Solve for θ in $3\tan \theta - 1 = 2$.

Solve It

20. Solve for θ in $\sin^2 \theta = \sin \theta$.

Solve It

21. Solve for θ in $2\cos^2 \theta - 1 = 0$.

Solve It

22. Solve for θ in $4\sin^2 \theta + 3 = 4$.

Solve It

23. Solve for θ in $4\sin^4 - 7\sin^2\theta + 3 = 0$.

Solve It

24. Solve for θ in $\tan^2\theta - \tan\theta = 0$.

Solve It

Making and Measuring Arcs

If someone asks you how far an ant on the edge of a 6-inch CD travels if the CD spins at 120°, you'd probably ask yourself, why do I care? You may even be thinking that the ant is probably messing up your CD player! But wacky math teachers love coming up with questions like that, so we're here to help you solve them.

To calculate the measure of an *arc*, a portion of the circumference of a circle like the path that pesky ant is taking, you need to remember that arcs can be measured in two ways: as an angle and as a length. As an angle, there's nothing to calculate — it's simply the same measure as the central angle. As a length, the measure of the arc is directly proportional to the circumference of the circle. Here's a nifty formula to give

you this value: $s = \dfrac{\theta}{360} \cdot 2\pi r$

Figure 6-5:
Calculating arc length and the variables involved.

EXAMPLE

Q. Let's find out about that ant! A pesky ant is on the edge of a 6-inch CD. How far does the ant travel if the CD spins at 120°?

A. **The ant travels approximately 6.3 inches.** First, remember that the angle needs to be in radians, so multiply the 120° by $\frac{\pi}{180°}$ to get $\frac{2\pi}{3}$. The diameter is 6 inches, so the radius is 3 inches. Use the formula $s = \theta \cdot r = \frac{2\pi}{3} \cdot 3 = 2\pi$ inches, which is approximately 6.3 inches.

---

25. Find the length of an arc in a circle with a radius of 4 feet if the central angle is $\frac{5}{6}$.

Solve It

26. Find the length of an arc in a circle with a diameter of 16 centimeters if the central angle is $\frac{7\pi}{4}$.

Solve It

27. Find the length of an arc in a circle with a radius of 18 feet if the angle is 210°.

Solve It

28. Find the radius of an arc in a circle that has an arc length of 42 inches if the angle is 2 radians.

Solve It

Answers to Problems

1 Find cosA in △ABC. **The answer is cosA = $\frac{5\sqrt{41}}{41}$.**

Because cosθ = $^{adj}/_{hyp}$, you need to find the adjacent side using the Pythagorean Theorem:
$x^2 + 4^2 = \left(\sqrt{41}\right)^2$, $x^2 + 16 = 41$, $x^2 = 25$, $x = 5$. Therefore, cosA = $\frac{5}{\sqrt{41}}$. Rationalizing the denominator, you get $\frac{5\sqrt{41}}{41}$.

2 Solve △DEF. **The answer is ∠E = 65°, side e = 23.6 cm, and side f = 26 cm.**

First, because you know ∠D and ∠F, you can find ∠E by subtracting ∠D and ∠F from 180°:
180° − (25° + 90°) = 65°. To find side e, you can use tan65° = $^e/_{11}$. Multiplying both sides by 11, you get e = 11 · tan65° = 23.6 cm. To find side f, you can use sin25° = $^{11}/_f$. Multiply both sides by f:
f · sin25° = 11. Divide by sin25° to get $f = \frac{11}{\sin 25°}$, which is approximately 26 cm.

3 Find ∠Q in △QRS (round to the nearest tenth). **The answer is approximately 22.6°.**

Because you have the adjacent side to ∠Q and the hypotenuse, you use cosine: cosQ = ($^{12}/_{13}$). To solve, take the inverse cosine of each side: Q = cos⁻¹($^{12}/_{13}$), which is approximately 22.6°.

4 Find the six trigonometric ratios of ∠R in △QRS from question 3. **The answer is sinR = $^{12}/_{13}$, cosR = $^5/_{13}$, tanR = $^{12}/_5$, cscR = $^{13}/_{12}$, secR = $^{13}/_5$, and cotR = $^5/_{12}$.**

Start by using the Pythagorean Theorem to find the third side: $12^2 + q^2 = 13^2$, $144 + q^2 = 169$, $q^2 = 25$, $q = 5$. Then, plug the sides into the trigonometric ratios: sinR = $^{12}/_{13}$, cosR = $^5/_{13}$, tanR = $^{12}/_5$, cscR = $^{13}/_{12}$, secR = $^{13}/_5$, cotR = $^5/_{12}$.

5 Romero wants to deliver a rose to his girlfriend, Jules, who is sitting on her balcony 24 feet above the street. If Romero has a 28-foot ladder, at what angle must he place the bottom of the ladder to reach his love, Jules? **The answer is 59°.**

To solve, draw a picture:

With the picture, you can see that you have the opposite side, 24 feet, from the angle you want and the hypotenuse, 28 feet. Therefore, you can use the sine ratio to solve: sinθ = $^{24}/_{28}$. To isolate the angle, use inverse sine: θ = sin⁻¹($^{24}/_{28}$), which is approximately 59°.

6   Sam needs to cross a river. He spies a bridge directly ahead of him. Looking across the river, he sees that he's 27° below the bridge from the other side. How far must he walk on his side of the river to reach the bridge if the bridge length is 40 feet? **The answer is 79 more feet.**

First, draw a picture:

Considering that you have the adjacent side from the angle, 40 feet, and you're looking for the opposite side, you can use the tangent ratio: $\tan 27° = \dfrac{40}{x}$. Multiplying both sides by x, you get $40 \cdot \tan 27° = x$. Dividing 40 by $\tan 27°$, you get that x equals approximately 79 feet.

7   Paul, a 6-foot-tall man, is holding the end of a kite string 5 feet above the ground. The string of the kite is 75-feet long at 35° of elevation. Paulette, Paul's 5-foot-tall daughter, is directly underneath the kite. How far above Paulette's head is the kite? **The answer is about 43 feet.**

Begin by (you guessed it!) drawing a picture:

Because Paul is holding the end of the kite string at the same height as Paulette's head, you only need to consider the string of the kite, which forms the hypotenuse of the triangle and the angle. Because you're looking for the opposite side from the angle and you have the hypotenuse, use the sine ratio to solve: $\sin 35° = x/75$. Multiplying both sides by 75, you get $x = \sin 35° \cdot 75$, which is approximately 43 feet.

8 To hold up a 100-foot pole, two guide wires are put up on opposite sides of the pole. One wire makes a 36° angle with the ground and the other makes a 47° angle with the ground. How far apart are the bases of the wires? **The answer is about 231 feet apart.**

Using your picture, you can set up two tangent ratios: $\tan 36° = {}^{100}/_x$ and $\tan 47° = {}^{100}/_y$. Multiply both sides by x and y respectively: $x \cdot \tan 36° = 100$ and $y \cdot \tan 47° = 100$. Isolate the variables

by dividing: $x = \dfrac{100}{\tan 36°}$ and $y = \dfrac{100}{\tan 47°}$. Therefore, x is approximately 137.6 feet and y is approximately 93.3. Add these together to get that the total distance apart is about 231 feet.

9 Find three co-terminal angles of $\pi/6$. **Although there are multiple answers, three possible answers are $^{11\pi}\!/_6$, $-^{9\pi}\!/_6$, and $^{21\pi}\!/_6$.**

Simply add or subtract multiples of 2π: $\pi/6 + 2\pi = {}^{11\pi}\!/_6$; $\pi/6 - 2\pi = -{}^{9\pi}\!/_6$; and $\pi/6 + 2 \cdot 2\pi = {}^{21\pi}\!/_6$.

10 Find two positive co-terminal angles of $-775°$. **Two possible answers are 305° and 665°.**

Here, just add multiples of 360° to $-775°$ until you get at two positive co-terminal angles: $-775° + 3 \cdot 360° = 305°$; $-775° + 4 \cdot 360° = 665°$.

11 Evaluate the six trigonometric ratios of the point $(3, 4)$. **$\sin\theta = {}^4\!/_5$, $\cos\theta = {}^3\!/_5$, $\tan\theta = {}^4\!/_3$, $\csc\theta = {}^5\!/_4$, $\sec\theta = {}^5\!/_3$, and $\cot\theta = {}^3\!/_4$.**

First, find the radius using the Pythagorean Theorem: $3^2 + 4^2 = r^2$, $9 + 16 = r^2$, $25 = r^2$, $5 = r$. Using this and $x = 3$, $y = 4$, plug into the trigonometric ratios: $\sin\theta = {}^4\!/_5$, $\cos\theta = {}^3\!/_5$, $\tan\theta = {}^4\!/_3$, $\csc\theta = {}^5\!/_4$, $\sec\theta = {}^5\!/_3$, $\cot\theta = {}^3\!/_4$.

12 Evaluate the six trigonometric ratios of the point $(-5, -7)$. **$\sin\theta = -\dfrac{7\sqrt{74}}{74}$, $\cos\theta = -\dfrac{5\sqrt{74}}{74}$, $\tan\theta = {}^7\!/_5$, $\csc\theta = -\dfrac{\sqrt{74}}{7}$, $\sec\theta = -\dfrac{\sqrt{74}}{5}$, and $\cot\theta = {}^5\!/_7$.**

Find your radius: $(-5)^2 + (-7)^2 = r^2$, $25 + 49 = r^2$, $74 = r^2$, $\sqrt{74} = r$. Using this and the point

(x, y), plug into the trig ratios and rationalize if necessary: $\sin\theta = \dfrac{-7}{\sqrt{74}} = -\dfrac{7\sqrt{74}}{74}$;

$\cos\theta = \dfrac{-5}{\sqrt{74}} = -\dfrac{5\sqrt{74}}{74}$; $\tan\theta = -{}^7\!/_{-5} = {}^7\!/_5$; $\csc\theta = -\dfrac{\sqrt{74}}{7}$; $\sec\theta = -\dfrac{\sqrt{74}}{5}$; and $\cot\theta = -{}^5\!/_{-7} = {}^5\!/_7$.

13 Evaluate the six trigonometric ratios of the point $(-2, 2\sqrt{3})$. $\sin\theta = \dfrac{\sqrt{3}}{2}$, $\cos\theta = -\frac{1}{2}$,

$\tan\theta = -\sqrt{3}$, $\csc\theta = 2\dfrac{\sqrt{3}}{3}$, $\sec\theta = -2$, and $\cot\theta = -\dfrac{\sqrt{3}}{3}$.

Begin by finding your radius: $(-2)^2 + \left(2\sqrt{3}\right)^2 = r^2$, $4 + 12 = r^2$, $16 = r^2$, $4 = r$. Now plug into your trig

ratios and rationalize if necessary: $\sin\theta = 2\dfrac{\sqrt{3}}{4} = \dfrac{\sqrt{3}}{2}$; $\cos\theta = -\frac{2}{4} = -\frac{1}{2}$; $\tan\theta = \dfrac{-2}{2\sqrt{3}} = -\sqrt{3}$;

$\csc\theta = \dfrac{4}{2\sqrt{3}} = \dfrac{2}{\sqrt{3}} = 2\dfrac{\sqrt{3}}{3}$; $\sec\theta = \frac{4}{-2} = -2$; and $\cot\theta = \dfrac{-2}{2\sqrt{3}} = -\dfrac{1}{\sqrt{3}} = -\dfrac{\sqrt{3}}{3}$.

14 Evaluate the six trigonometric ratios of the point $(6, -3\sqrt{5})$. $\sin\theta = -\dfrac{\sqrt{5}}{3}$, $\cos\theta = \frac{2}{3}$,

$\tan\theta = -\dfrac{\sqrt{5}}{2}$, $\csc\theta = -3\dfrac{\sqrt{5}}{5}$, $\sec\theta = \frac{3}{2}$, and $\cot\theta = -2\dfrac{\sqrt{5}}{5}$.

Start by finding the radius: $6^2 + \left(-3\sqrt{5}\right)^2 = r^2$, $36 + 45 = r^2$, $81 = r^2$, $9 = r$. Plug into your trig ratios

and rationalize if necessary: $\sin\theta = -3\dfrac{\sqrt{5}}{9} = -\dfrac{\sqrt{5}}{3}$; $\cos\theta = \frac{6}{9} = \frac{2}{3}$; $\tan\theta = -3\dfrac{\sqrt{5}}{6} = -\dfrac{\sqrt{5}}{2}$;

$\csc\theta = \dfrac{9}{-3\sqrt{5}} = -\dfrac{3}{\sqrt{5}} = -3\dfrac{\sqrt{5}}{5}$; $\sec\theta = \frac{9}{6} = \frac{3}{2}$; and $\cot\theta = \dfrac{6}{-3\sqrt{5}} = -\dfrac{2}{\sqrt{5}} = -2\dfrac{\sqrt{5}}{5}$.

15 Evaluate the six trigonometric ratios of 225°. $\sin\theta = -\dfrac{\sqrt{2}}{2}$, $\cos\theta = -\dfrac{\sqrt{2}}{2}$, $\tan\theta = 1$,

$\csc\theta = -\sqrt{2}$, $\sec\theta = -\sqrt{2}$, and $\cot\theta = 1$.

Using reference angles, you can see that you're dealing with a 45-45-90 triangle (225° − 180°).

Therefore, $x = -\dfrac{\sqrt{2}}{2}$ and $y = -\dfrac{\sqrt{2}}{2}$. Now, by using the point-in-plane definition, you can find

the sixtrig ratios. $\sin\theta = x = -\dfrac{\sqrt{2}}{2}$; $\cos\theta = y = -\dfrac{\sqrt{2}}{2}$; $\tan\theta = \frac{y}{x} = -\dfrac{\sqrt{2}}{2} \div -\dfrac{\sqrt{2}}{2} = 1$;

$\csc\theta = \frac{1}{x} = 1 \div -\dfrac{\sqrt{2}}{2} = -\sqrt{2}$; $\sec\theta = \frac{1}{y} = 1 \div -\dfrac{\sqrt{2}}{2} = -\sqrt{2}$; and $\cot\theta = \frac{x}{y} = \frac{y}{x} = -\dfrac{\sqrt{2}}{2} \div -\dfrac{\sqrt{2}}{2} = 1$.

16 Find θ when $\cos\theta = \dfrac{\sqrt{3}}{2}$ and $0° < \theta < 360°$. **The answer is 60° and 300°.**

Looking at the special right triangles, you can see that $\cos\theta = \dfrac{\sqrt{3}}{2}$ when θ is 60°. Because

cosine is equal to the x value on the unit circle and x is positive in quadrants I and IV, the answer is 60° and 300°.

17 Evaluate the six trigonometric ratios of 330°. $\sin\theta = -\frac{1}{2}$, $\cos\theta = \frac{\sqrt{3}}{2}$, $\tan\theta = -\frac{\sqrt{3}}{3}$, $\csc\theta = -2$, $\sec\theta = \frac{\sqrt{3}}{3}$, **and** $\cot\theta = -\sqrt{3}$.

Considering that 330° is in quadrant IV, using reference angles (360°–330°), you find that you're dealing with a 30-60-90 triangle. Using the point-in-plane definition, you get $\sin\theta = x = -1/2$;

$\sqrt{\cos\theta} = y = \frac{\sqrt{3}}{2}$; $\tan\theta = \frac{y}{x} = -\frac{1}{2} \div \frac{\sqrt{3}}{2} = -\frac{1}{2} \cdot \frac{2}{\sqrt{3}} = -\frac{1}{\sqrt{3}} = -\frac{\sqrt{3}}{3}$; $\csc\theta = \frac{1}{y} = 1 \div -\frac{1}{2} = -2$; $\sqrt{\sec\theta} = \frac{1}{x} = 1 \div \frac{\sqrt{3}}{2} = \frac{2}{\sqrt{3}} = \frac{2\sqrt{3}}{3}$; and $\cot\theta = \frac{x}{y} = \frac{\sqrt{3}}{2} \div +\frac{1}{2} = \frac{\sqrt{3}}{2} \cdot \frac{2}{1} = -\sqrt{3}$.

18 Find θ when $\tan\theta = \frac{\sqrt{3}}{3}$ and $0° < \theta < 360°$. **The answer is $\theta = 30°$ and $210°$.**

To solve this, use special right triangles. You can see that if θ is 30°, then $x = \frac{1}{2}$ and $y = \frac{\sqrt{3}}{2}$.

This means that because $\tan\theta = \frac{y}{x}$, $\tan\theta = \frac{1}{2} \div \frac{\sqrt{3}}{2} = \frac{1}{2} \cdot \frac{2}{\sqrt{3}} = \frac{1}{\sqrt{3}} = \frac{\sqrt{3}}{3}$, which is what you want.

Therefore, $\theta = 30°$ and $210°$.

19 Solve for θ in $3\tan\theta - 1 = 2$. **The answer is $\theta = 45°$ and $225°$.**

Begin by using algebra to isolate $\tan\theta$: $3\tan\theta - 1 = 2$, $3\tan\theta = 3$, $\tan\theta = 1$. Because $\tan\theta = y/x$, sine and cosine must be the same value for $\tan\theta$ to equal 1. This occurs at $\theta' = 45°$. Because the answer is positive, both sine and cosine must be the same sign, which occurs in quadrants I and III. Therefore, using reference angles, quadrant I: $\theta = 45°$, and quadrant III: $180° + 45° = 225°$.

20 Solve for θ in $\sin^2\theta = \sin\theta$. **The answer is $\theta = 0$, $\frac{1}{2}$, π, and 2π.**

To solve, think of $\sin^2\theta = \sin\theta$ as $x^2 = x$, which can be solved by bringing both terms to the same side and factoring: $x^2 - x = 0$, $x(x - 1) = 0$. Similarly, $\sin^2\theta = \sin\theta$, $\sin^2\theta - \sin\theta = 0$, $\sin\theta(\sin\theta - 1) = 0$. Therefore, $\sin\theta = 0$, or $\sin\theta - 1 = 0$, which means $\sin\theta = 1$. Knowing that $\sin\theta = y$ on the unit circle, $\sin\theta = 0$ at 0, and π, 2π and $\sin\theta = 1$ at $\frac{1}{2}$, you have your answers!

21 Solve for θ in $2\cos^2\theta - 1 = 0$. **The answer is $\theta = \frac{\pi}{4}$, $\frac{3\pi}{4}$, $\frac{5\pi}{4}$, and $\frac{7\pi}{4}$.**

First, isolate the cosine term using algebra: $2\cos^2\theta - 1 = 0$, $2\cos^2\theta = 1$, $\cos^2\theta = \frac{1}{2}$. Now, take the square root of each side: $\sqrt{\cos^2\theta} = \pm\sqrt{\frac{1}{2}}$. Thus, $\cos\theta = \pm\frac{1}{\sqrt{2}} = \pm\frac{\sqrt{2}}{2}$. This occurs at four angles on the unit circle: $\frac{\pi}{4}$, $\frac{3\pi}{4}$, $\frac{5\pi}{4}$, and $\frac{7\pi}{4}$.

22 Solve for θ in $4\sin^2\theta + 3 = 4$. **The answer is $\theta = \frac{\pi}{6}$, $\frac{5\pi}{6}$, $\frac{7\pi}{6}$, and $\frac{11\pi}{6}$.**

Begin by using algebra to isolate the sine term: $4\sin^2\theta + 3 = 4$, $4\sin^2\theta = 1$, $\sin^2\theta = \frac{1}{4}$. Taking the square root of each side, you get: $\sqrt{\sin^2\theta} = \pm\sqrt{\frac{1}{4}}$, $\sin\theta \pm\frac{1}{2}$. This means that $\theta = \frac{\pi}{6}$, $\frac{5\pi}{6}$, $\frac{7\pi}{6}$, and $\frac{11\pi}{6}$.

23 Solve for θ in $4\sin^4\theta - 7\sin^2\theta + 3 = 0$ in degrees. **The answer is $\theta = 0°$, $60°$, $120°$, $240°$, $300°$, $90°$, $180°$, $360°$, and $270°$.**

Start by thinking of $4\sin^4\theta - 7\sin^2\theta + 3 = 0$ as $4x^4 - 7x^2 + 3 = 0$, which factors into $(4x^2 - 3)(x^2 - 1) = 0$. Similarly, $4\sin^4\theta - 7\sin^2\theta + 3 = 0$ factors into $(4\sin^2\theta - 3)(\sin^2\theta - 1) = 0$. Set each factor equal to zero and take the square root of each side to find $\sin\theta$: $4\sin^2\theta - 3 = 0$, $4\sin^2\theta = 3$, $\sin^2\theta = \frac{3}{4}$,

$\sqrt{\sin^2\theta} = \pm\sqrt{\frac{3}{4}}$, $\sin\theta = \pm\frac{\sqrt{3}}{2}$. Therefore, $\theta = 60°$, $120°$, $240°$, $300°$. Or, $\sin^2\theta - 1 = 0$, $\sin^2\theta = 1$,

$\sqrt{\sin^2\theta} = \pm\sqrt{1}$, $\sin\theta = \pm 1$. Therefore, $\sin\theta$ also equals $0°$, $90°$, $180°$, $360°$, and $270°$.

24 Solve for θ in $\tan^2\theta - \tan\theta = 0$. **The answer is θ = 0, π, 2π, ¼, and ⁵π⁄₄.**

Notice that this one is similar to question 20. You can factor the same way: $\tan^2\theta - \tan\theta = 0$, $\tan\theta(\tan\theta - 1) = 0$. Setting each factor to zero: $\tan\theta = 0$ or $\tan\theta - 1 = 0$, so $\tan\theta = 1$. These occur at 0, π, 2π, ¼, and ⁵π⁄₄.

25 Find the length of an arc in a circle with a radius of 4 feet if the central angle is ⁵⁄₆. **The answer is s = ²⁰⁄₃ feet, or approximately 2.1 feet.**

Use the formula $s = \theta \cdot r$ = ⁵⁄₆ · 4 = ²⁰⁄₃ feet, which is approximately 2.1 feet.

26 Find the length of an arc in a circle with a diameter of 16 centimeters if the central angle is ⁷π⁄₄. **The answer is s = 14π cm ≈ 44cm.**

Start by finding the radius by dividing the diameter by two: 8 cm. Next, plug into the arc length formula: $s = \theta \cdot r$ = ⁷π⁄₄ · 8 = 14π cm ≈ 44cm.

27 Find the length of an arc in a circle with a radius of 18 feet if the angle is 210°. **The answer is s = 21π feet, which is approximately 66 feet.**

Begin by changing degrees to radians by multiplying 210° by ⁷π⁄₁₈₀: 210° · ⁷π⁄₁₈₀ = ⁷π⁄₆. Now, plug in the radius and angle into the arc length formula: $s = \theta \cdot r$ = ⁷π⁄₆ · 18 = 21π feet, which is approximately 66 feet.

28 Find the radius of an arc in a circle that has an arc length of 42 inches if the angle is 2 radians. The answer is 21 inches.

Just plug 'em in! $s = \theta \cdot r$, 42 = 2 · r. Dividing both sides by 2, you find r = 21 inches.

Chapter 7

Graphing and Transforming Trig Functions

In This Chapter
▶ Exploring period graphs
▶ Graphing sine and cosine
▶ Picturing tangent and cotangent
▶ Charting secant and cosecant

Graphing trig functions is similar to graphing any other function. You simply insert values into the input to find the output. In this case, the input is typically θ and the output is typically *y*. And, just like graphing any other function, knowing the *parent trig graph* — the most basic, unshifted graph — makes graphing more complex graphs easier. In this chapter, we show you the parent graph of each trig function and its transformations.

Getting a Grip on Period Graphs

Periodic graphs are like that everlasting bunny we all know . . . they keep going, and going, and going. Well, when graphing trig functions, if you remember that they're periodic graphs, then the steps to graphing them are easy! Because they repeat their values over and over again, you just need to figure out one period (or cycle), and you can repeat it as many times as you like.

The key is to graph one period. To do this, you need to start by graphing the parent graph, then make transformations (just like you do in Chapter 3 for other types of graphs). Like other graphs, trig graphs make vertical and horizontal transformations, as well as vertical and horizontal translations:

✔ Presto, chango! Vertical and horizontal transformations:

- For trig functions, vertical transformations are achieved by changing the *amplitude,* or height. To do this, simply multiply a constant by the parent function. For example, $f(\pi) = 2\sin\pi$ is the same as the parent graph, only its wave goes up to a value of 2 and down to −2.

- Multiplying a negative constant to the parent graph simply flips the graph upside down, or reflects it over the *x*-axis.

- Horizontal transformations occur by changing the period of the graph. For sine and cosine parent graphs, the period is 2π. The same is true for cosecant and secant graphs. For tangent and cotangent, the period is π. Multiplying a constant by the angle transforms the period, or makes a horizontal transformation. For example, $f(\theta) = \cos 2\theta$ results in a graph that repeats itself twice in the amount of space the parent graph would.

✓ Vertical and horizontal translations — shifting the parent graph up, down, left, or right:

- Just as we show you in Chapter 3, vertical and horizontal shifts just change the location on the graph: up, down, left, or right.

- The general equation for these shifts for sine, for example, is $\sin(\theta - h) + v$, where h represents the horizontal shift left or right and v represents the vertical shift up or down.

- To find the horizontal shift of a function, simply set the inside parentheses equal to 0. For example, the horizontal shift for $\sin(\theta + 3)$ is -3 because $\theta + 3 = 0$, so $\theta = -3$.

We suggest that when you put the transformations together, you follow this simple order:

1. **Change the amplitude.**

2. **Change the period.**

3. **Shift the graph horizontally.**

4. **Shift the graph vertically.**

Sine and Cosine: Parent Graphs and Transformations

Sine and cosine graphs look like waves. These waves, or *sinusoids* in math speak, keep going and going like our bunny friend. To graph these sinusoids, you need to start with checking out the parent graphs (see Figure 7-1).

Figure 7-1: Parent graphs of sine and cosine.

$y = \sin\theta$

$y = \cos\theta$

Notice that the periods of both the sine and cosine graphs are the same: 2π. Similarly, they both have an amplitude (or height) of 1. You use this information for your transformations.

Putting together all the transformation information into one equation, you get

$$f(\theta) = a \cdot \sin[p(\theta - h)] + v$$

$$f(\theta) = a \cdot \cos[p(\theta - h)] + v$$

where a is the amplitude, h is the horizontal shift, v is the vertical shift, and you divide 2π by p to get the period.

Q. Graph $2\sin\theta + 3$.

A. Starting with amplitude, you can see that $a = 2$, so your amplitude is 2. The period is

2π because the period doesn't change from the parent equation. The vertical shift is positive 3 because $v = 3$. The graph is shown in the following figure.

---

1. Graph $f(\theta) = \dfrac{-1}{2} \cdot \cos\theta$.

Solve It

2. Graph $f(\theta) = \cos\dfrac{1}{2}\theta$.

Solve It

3. Graph $f(\theta) = \sin(\theta + \frac{3}{4})$.

Solve It

4. Graph $f(\theta) = \cos\frac{1}{3}\theta + 2$.

Solve It

5. Name the amplitude, period, horizontal shift, and vertical shift of $f(\theta) = 3\sin(2\theta + \frac{3}{2}) - 1$.

Solve It

6. Graph $f(\theta) = 3\sin(2\theta + \frac{3}{2}) - 1$.

Solve It

Tangent and Cotangent: Mom, Pops, and Children

Tangent and cotangent are both periodic, but they're not wavelike like sine and cosine. Instead, they have vertical asymptotes that break up their graphs. As we discuss in Chapter 3, a *vertical asymptote* is where the function is undefined. Because tangent and cotangent are rational functions where $\tan\theta = \dfrac{\sin\theta}{\cos\theta}$ and $\cot\theta = \dfrac{\cos\theta}{\sin\theta}$, they both have values that are undefined where their denominators are equal to 0. For tangent, this occurs on the unit circle at $\frac{1}{2}$ and $\frac{3}{2}$. For cotangent, this occurs at 0, π, and 2π on the unit circle. Therefore, these are the locations of their asymptotes (see Figure 7-2).

Figure 7-2: Parent graphs of tangent and cotangent.

$y = \tan\theta$

$y = \cot\theta$

Notice that the periods of both the tangent and cotangent graphs are the same: π. The *x*-intercepts for tangent are 0, π, and 2π. For cotangent, the *x*-intercepts are $\frac{1}{2}$ and $\frac{3}{2}$. Using this information, you can make your transformations.

Putting together all the transformation information from earlier in this chapter into one equation, you get

$$f(\theta) = a \cdot \tan[p(\theta - h)] + v$$

$$f(\theta) = a \cdot \cot[p(\theta - h)] + v$$

where a is the vertical transformation (no amplitude with tangent and cotangent), h is the horizontal shift, v is the vertical shift, and you divide π by p to get the period.

Q. Graph $f(\theta) = \frac{1}{2} \cdot \tan 2\theta$.

A. Starting with the vertical transformation, you can see that it's ½. Next, find the period by dividing π by 2, which is ½. Because there are no horizontal or vertical shifts, you're ready to graph:

7. Graph $f(\theta) = \cot \frac{1}{2}\theta$.

Solve It

8. Graph $f(\theta) = \tan\theta + 2$.

Solve It

9. Graph $f(\theta) = \frac{1}{3} \cdot \cot\theta$.

Solve It

10. Graph $f(\theta) = \tan(\theta - \frac{1}{2})$.

Solve It

11. Name the amplitude, period, horizontal shift, and vertical shift of $f(\theta) = 2 \cdot \tan(\theta + \frac{3}{4}) - 1$.

Solve It

12. Graph $f(\theta) = 2 \cdot \tan(\theta + \frac{3}{4}) - 1$.

Solve It

Secant and Cosecant: Generations

To graph cosecant and secant, it's important to remember that they're the reciprocals of sine and cosine, respectively: $\csc\theta = \dfrac{1}{\sin\theta}$ and $\sec\theta = \dfrac{1}{\cos\theta}$. Using this fact, the easiest way to graph cosecant or secant is to start by graphing sine or cosine — the graphs of the reciprocals are easily found from there!

Given that cosecant and secant are reciprocal functions of sine and cosine, this also means that they're rational functions. And just as with tangent and cotangent, rational functions often mean asymptotes. So, yep, you guessed it — cosecant and secant graphs have asymptotes. These occur wherever their reciprocal functions (sine or cosine) have a value of 0. To graph, follow these easy steps:

1. **Graph the sine graph with transformations to graph a cosecant graph, or graph the cosine graph with transformations to get the secant graph.**

2. **Draw asymptotes where the sine or cosine functions are equal to 0.**

3. **Sketch the reciprocal graph of cosecant or secant between each pair of asymptotes.**

    For example, if the sine graph gets bigger, the cosecant graph would get smaller.

The parent graphs of cosecant and secant are shown in Figure 7-3. The parent sine and cosine graphs are also left for you to see where they came from.

$y = \csc\theta$ $y = \sec\theta$

Figure 7-3:
Parent
graphs of
cosecant
and secant.

Q. Graph $f(\theta) = \csc\theta + 1$.

A. Here there's only a vertical shift of 1, which means you shift the parent graph up one (see the following figure).

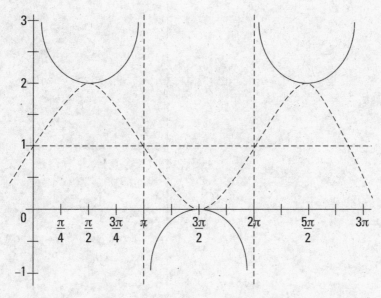

---

13. Graph $f(\theta) = -\csc\theta - 1$.

Solve It

14. Graph $f(\theta) = \sec 2\theta + 1$.

Solve It

15. Name the asymptotes from 0 to 2π, and the horizontal and vertical shifts of

$$f(\theta) = \frac{1}{4} \cdot \csc\left(\theta - \frac{\pi}{2}\right) - 1 .$$

Solve It

16. Graph $f(\theta) = \frac{1}{4} \cdot \csc\left(\theta - \frac{\pi}{2}\right) - 1 .$

Solve It

17. Name the amplitude, period, horizontal shift, and vertical shift of

$$f(\theta) = 2 \cdot \sec\frac{1}{2} \cdot \left(\theta + \frac{\pi}{4}\right) + 1 .$$

Solve It

18. Graph $f(\theta) = 2 \cdot \sec\frac{1}{2} \cdot \left(\theta + \frac{\pi}{4}\right) + 1 .$

Solve It

Answers to Problems on Graphing and Transforming Trig Functions

1 Graph $f(\theta) = -\frac{1}{2} \cdot \cos\theta$. **See the following figure.**

Because the cosine function is multiplied by $-\frac{1}{2}$, the graph is inverted with an amplitude of $\frac{1}{2}$. The period doesn't change, and there are no shifts.

2 Graph $f(\theta) = \cos\frac{1}{2} \cdot \theta$. **See the following figure.**

Here, the amplitude doesn't change, but the period does. The new period is found by dividing 2π by $\frac{1}{2}$, which is 4π. There are no vertical or horizontal shifts.

3 Graph $f(\theta) = \sin(\theta + \frac{3}{4})$. **See the following figure.**

This graph has a horizontal shift. To find it, set what's inside the parentheses to the starting value of the parent graph: $\theta + \frac{3}{4} = 0$, so $\theta = -\frac{3}{4}$. There are no other changes from the parent graph.

4 Graph $f(\theta) = \cos\frac{1}{3} \cdot \theta + 2$. **See the following figure.**

This has a change in period, which can be found by dividing 2π by $\frac{1}{3}$ to get 6π. There's also a vertical shift of 2.

5 Name the amplitude, period, horizontal shift, and vertical shift of $f(\theta) = 3\sin[2(\theta + \frac{5}{2})] - 1$. **The amplitude is 3, the period is π, the horizontal shift is $-\frac{5}{2}$, and the vertical shift is -1.**

The only calculation you need to do is to find the period. Here, you divide 2π by 2 to get π. From the equation, you can see that the amplitude is 3 and the vertical shift is -1. By setting $\theta + \frac{5}{2} = 0$, you find that the horizontal shift is $-\frac{5}{2}$.

6 Graph $f(\theta) = 3\sin(2\theta + \frac{5}{2}) - 1$. **See the following figure.**

Using the information you gather in problem 5, this graph comes together quickly.

7   Graph $f(\theta) = \cot\frac{1}{2}\theta$. **See the following figure.**

For this cotangent graph, the period has a change, which can be found by dividing π by $\frac{1}{2}$ (you get 2π). There are no other changes to the parent cotangent graph.

8   Graph $f(\theta) = \tan\theta + 2$. **See the following figure.**

Here, there's only a vertical shift of 2. The amplitude and period are the same as the parent graph.

9 Graph $f(\theta) = \frac{1}{3} \cdot \cot\theta$. **See the following figure.**

This graph shows a vertical transformation of $\frac{1}{3}$.

10 Graph $f(\theta) = \tan(\theta - \frac{\pi}{2})$. **See the following figure.**

This tangent graph has a horizontal shift of $\frac{\pi}{2}$. There are no other changes to the parent graph.

11 Name the vertical transformation, period, horizontal shift, and vertical shift of $f(\theta) = 2 \cdot \tan(\theta + \frac{3}{4}) - 1$. **The vertical transformation is 2, the period is π, the horizontal shift is $-\frac{3}{4}$, and the vertical shift is -1.**

The only calculation here is to find the horizontal shift by setting $\theta + \frac{3}{4} = 0$. You get $\theta = -\frac{3}{4}$. You get the rest of the info straight from the equation.

12 Graph $f(\theta) = 2 \cdot \tan(\theta + \frac{\pi}{4}) - 1$. **See the following figure.**

Using the information you gather in problem 11, you get the following graph:

13 Graph $f(\theta) = -\csc\theta - 1$. **See the following figure.**

REMEMBER

For these you start by graphing the sine graph to find the asymptotes, and then sketch the reciprocal function of cosecant. We include the sine graph for you to see that it has been flipped because of the negative in front and shifted down 1 because of the vertical shift.

14 Graph $f(\theta) = \sec 2\theta + 1$. **See the following figure.**

Again, we include the reciprocal cosine graph. This has a change in period, which you find by dividing 2π by 2 to get π. It also has a vertical shift of 1. Then you draw the asymptotes and sketch the secant graph.

15 Name the asymptotes from 0 to 2π, and the horizontal and vertical shifts of $f(\theta) = \frac{1}{4}\csc(\theta - \frac{\pi}{2}) - 1$. **The asymptotes are at $\frac{\pi}{2}$, $\frac{3\pi}{2}$, and $\frac{5\pi}{2}$. The horizontal shift is $-\frac{\pi}{2}$, and the vertical shift is -1.**

To find the asymptotes, you need to first look for any shifts or changes in period that would affect the parent graph. The shifts are evident from the equation, where the horizontal shift is $-\frac{\pi}{2}$ and the vertical shift is -1. This makes the zeros of the reciprocal sine graph at $\frac{\pi}{2}$, $\frac{3\pi}{2}$, and $\frac{5\pi}{2}$. This is then where the asymptotes will be.

16 Graph $f(\theta) = \frac{1}{4}\csc(\theta - \frac{\pi}{2}) - 1$. **See the following figure.**

Using the information from problem 15, sketch the sine graph, the asymptotes, and then the reciprocal — the cosecant graph.

17 Name the amplitude, period, horizontal shift, and vertical shift of $f(\theta) = 2\sec\frac{1}{2}(\theta + \frac{\pi}{4}) + 1$. **The amplitude is 2, the period is 4π, the horizontal shift is $-\frac{\pi}{4}$, and the vertical shift is 1.**

The amplitude and shifts are evident from the equation. To find the period, simply divide 2π by $\frac{1}{2}$ and you get 4π.

18 Graph $f(\theta) = 2\sec(\frac{1}{2} \cdot \theta + \frac{\pi}{4}) + 1$. See the following figure.

Begin by using the information from problem 17 to sketch the reciprocal, the cosine graph. From there, sketch the asymptotes where the cosine graph equals 1 (or the cosine function equals 0). Then, sketch the reciprocal function to graph the secant graph.

Part III
Advanced Trig: Identities, Theorems, and Applications

The 5th Wave   By Rich Tennant

Sorry. I just find rotating my head helps me to relax during the test.

In this part . . .

The concepts of trig keep building, and we're here to help you follow along.

These chapters move into identities — they're like formulas but they're true all the time, no matter what you put in for the variable(s). These identities are used to simplify expressions and solve equations, and they're even used in trig proofs (and you thought you were done with proofs in geometry!).

Chapter 8

Basic Trig Identities

● ●

In This Chapter

▶ Simplifying with reciprocal identities

▶ Reducing using Pythagorean identities

▶ Finding patterns with even-odd identities

▶ Using co-function identities

▶ Discovering periodicity identities

▶ Solving trigonometric proofs

● ●

*E*ver want to pretend you were someone else — change your identity? Well, trig expressions have the opportunity to do that all the time. In this chapter, we cover basic *identities,* or statements that are always true. We use these identities to simplify problems and then to complete trigonometric proofs. Each section builds upon the previous one, so we recommend you spend some time reviewing the identities in each section before jumping to the end to practice proofs.

Using Reciprocal Identities to Simplify Trig Expressions

We actually introduce some of these back in Chapter 6, but now we're going to use reciprocal identities to simplify more complicated trig expressions. Because these identities are all review, we also include the ratios of tangent and cotangent — the *ratio identities* that we introduce in Chapter 6. The reciprocal (and ratio) identities are:

$$\sin x = \frac{1}{\csc x} \qquad \csc x = \frac{1}{\sin x}$$

$$\cos x = \frac{1}{\sec x} \qquad \sec x = \frac{1}{\cos x}$$

$$\tan x = \frac{1}{\cot x} = \frac{\sin x}{\cos x} \qquad \cot x = \frac{1}{\tan x} = \frac{\cos x}{\sin x}$$

Because each pair of expressions is mathematically equivalent, you can substitute one for another in a given expression and watch things simplify. Typically, changing a given expression to all sines and cosines causes a whole lot of canceling! Try it and see . . . we dare you!

Q. Use reciprocal identities to rewrite $\dfrac{\sin x \cdot \sec x}{\tan x}$.

A. **The answer is 1.** Start by using reciprocal and ratio identities to rewrite $\sec x$ and $\tan x$ in terms of sine and cosine (we dared you). Next, use your knowledge of fractions to rewrite the expression as a division problem. Then, multiply by the reciprocal and cancel where you can. Here's what it should look like:

$$\frac{\sin x \cdot \sec x}{\tan x} = \frac{\sin x \cdot \dfrac{1}{\cos x}}{\dfrac{\sin x}{\cos x}} = \sin x \cdot \frac{1}{\cos x} \div \frac{\sin x}{\cos x} = \sin x \cdot \frac{1}{\cos x} \cdot \frac{\cos x}{\sin x} = \cancel{\sin x} \cdot \frac{1}{\cancel{\cos x}} \cdot \frac{\cancel{\cos x}}{\cancel{\sin x}} = 1$$

1. Simplify $\cot x \cdot \sec x$.

Solve It

2. Simplify $\sin x \cdot \sec x$.

Solve It

3. Simplify $\sin^3 x \cdot \csc^2 x + \tan x \cdot \cos x$.

Solve It

4. Simplify $\cot x \cdot \sin x \cdot \tan x$.

Solve It

Simplifying with Pythagorean Identities

Pythagorean identities are extremely helpful for simplifying complex trig expressions. These identities are derived from those right triangles on a unit circle (turn to Chapter 6 for a review if you need to). Remember that $\cos\theta$ = the x leg of a triangle, $\sin\theta$ = the y leg of a triangle, and the hypotenuse of the triangle on that unit circle is 1. Given the fact that leg^2 + leg^2 = hypotenuse^2, we get the first Pythagorean identity. The other two are derived from that (check out *Pre-Calculus For Dummies* if you want to see how this works!). These identities are especially helpful when simplifying expressions that have a term that has been squared ($\sin^2$, $\cos^2$, and so on). Here are the Pythagorean identities (and some derivatives):

$$\sin^2 x + \cos^2 x = 1 \quad \text{or} \quad \cos^2 x = 1 - \sin^2 x \quad \text{or} \quad \sin^2 x = 1 - \cos^2 x$$

$$\tan^2 x + 1 = \sec^2 x \quad \text{or} \quad \tan^2 x = \sec^2 x - 1 \quad \text{or} \quad 1 = \sec^2 x - \tan^2 x$$

$$1 + \cot^2 x = \csc^2 x \quad \text{or} \quad \cot^2 x = \csc^2 x - 1 \quad \text{or} \quad 1 = \csc^2 x - \cot^2 x$$

$Q.$ Simplify $(\sec x + \tan x)(1 - \sin x)(\cos x)$.

$A.$ **$\cos^2 x$.** Start by changing everything to sine and cosine using the reciprocal and ratio identities from the previous section. Then add the resulting fractions (the common denominator is cosine) and cancel the cosine in the numerator and denominator. This leaves you with two terms that you can FOIL. Recognize this last term as a Pythagorean identity? We hoped you might! Substitute it in and you have your answer.

The steps look like this:

$$(\sec x + \tan x)(1 - \sin x)(\cos x) =$$

$$\left(\frac{1}{\cos x} + \frac{\sin x}{\cos x} \right)(1 - \sin x)(\cos x) =$$

$$\left(\frac{1 + \sin x}{\cos x} \right)(1 - \sin x)(\cos x) =$$

$$\left(\frac{1 + \sin x}{\cancel{\cos x}} \right)(1 - \sin x)(\cancel{\cos x}) =$$

$$(1 + \sin x)(1 - \sin x) =$$

$$1 - \sin^2 x =$$

$$\cos^2 x$$

---

5. Simplify $\dfrac{\cos x}{\sin x}(\tan x + \cot x)$.

Solve It

6. Simplify $1 - \dfrac{\sin x \cdot \tan x}{\sec x}$.

Solve It

---

7. Simplify $\sin x \cdot \cot^2 x + \sin x$.

Solve It

8. Simplify $(\sin^2 x - 1)(\tan^2 x + 1)$.

Solve It

Discovering Even-Odd Identities

All functions, including trig functions, can be described as being even, odd, or neither (see Chapter 3 for review). Knowing whether a trig function is even or odd can actually help you simplify an expression. These even-odd identities are helpful when you have an expression where the variable inside the trig function is negative (such as $-x$). The even-odd identities are:

$$\sin(-x) = -\sin x \qquad \csc(-x) = -\csc x$$
$$\cos(-x) = \cos x \qquad \sec(-x) = \sec x$$
$$\tan(-x) = -\tan x \qquad \cot(-x) = -\cot x$$

Q. Simplify $\tan^2(-x) + \sec^2(-x)$.

A. **The answer is 1.** Using the even-odd identities, start by substituting for the negative angles: $-\tan^2 x + \sec^2 x$. Using commutative property of equality (from Chapter 1), rewrite the expression: $\sec^2 x - \tan^2 x$. Recognize this from our Pythagorean identities from the last section? This expression equals 1.

9. Simplify $\sec(-x) \cdot \cot(-x)$.

Solve It

10. Simplify $\sin x \cdot [\csc x + \sin(-x)]$.

Solve It

11. Simplify $\dfrac{\csc x}{\cot(-x)}$.

Solve It

12. Simplify $\dfrac{\cot^2(-x) - 1}{\cot^2 x}$.

Solve It

Solving with Co-Function Identities

Ever notice that the graphs of sine and cosine look exactly alike, only shifted (see Chapter 7 for a visual)? This is also the case for tangent and cotangent, as well as for secant and cosecant. Because these functions have the same values, only shifted, we can define them as being *co-functions*. We can write them as co-function identities and use them to simplify expressions. The co-function identities are:

$$\sin x = \cos(\tfrac{\pi}{2} - x) \qquad\qquad \cos x = \sin(\tfrac{\pi}{2} - x)$$

$$\tan x = \cot(\tfrac{\pi}{2} - x) \qquad\qquad \cot x = \tan(\tfrac{\pi}{2} - x)$$

$$\csc x = \sec(\tfrac{\pi}{2} - x) \qquad\qquad \sec x = \csc(\tfrac{\pi}{2} - x)$$

Q. Simplify $\dfrac{\cot(\tfrac{\pi}{2} - x)}{\sec x}$.

A. **sin x.** Start by using the co-function identity to replace cot $(\tfrac{\pi}{2} - x)$ with tan x. Next, rewrite the fraction as a division problem. Then, rewrite in terms of sine and cosine using reciprocal and ratio identities. Finally, simplify by using the multiplicative inverse, cancelling any common terms. The steps look like this:

$$\frac{\cot(\tfrac{\pi}{2} - x)}{\sec x} = \frac{\tan x}{\sec x} = \tan x \div \sec x = \frac{\sin x}{\cos x} \div \frac{1}{\cos x} = \frac{\sin x}{\cancel{\cos x}} \cdot \frac{\cancel{\cos x}}{1} = \sin x$$

---

13. Simplify $\sin\left(\dfrac{\pi}{2} - x\right) \cdot \cot(\pi - x)$.

Solve It

14. Simplify
$$\sin\left(\frac{\pi}{2} - x\right) + \cot\left(\frac{\pi}{2} - x\right) \cdot \cos\left(\frac{\pi}{2} - x\right).$$

Solve It

---

15. Simplify $\dfrac{\cos(\tfrac{\pi}{2} - x)}{\sin(\tfrac{\pi}{2} - x)} \cdot \cot(\tfrac{\pi}{2} - x)$.

Solve It

16. Simplify
$$\left[\sec\left(\frac{\pi}{2} - x\right) + \tan\left(\frac{\pi}{2} - x\right)\right]\left[1 - \sin\left(\frac{\pi}{2} - x\right)\right].$$

Solve It

Moving with Periodicity Identities

Recall that horizontal transformations change the period of a graph or move it left or right (see Chapter 7). If you shift the graph by one whole period to the left or right, you end up with the same function. This is the idea behind *periodicity identities*. Because the periods of sine, cosine, cosecant, and secant repeat every 2π, and tangent and cotangent repeat every π, the periodicity identities are as follows:

$$\sin(x + 2\pi) = \sin x$$
$$\cos(x + 2\pi) = \cos x$$
$$\tan(x + \pi) = \tan x$$
$$\cot(x + \pi) = \cot x$$
$$\csc(x + 2\pi) = \csc x$$
$$\sec(x + 2\pi) = \sec x$$

EXAMPLE

Q. Simplify $1 - [\sin(2\pi + x) \cdot \cot(\pi + x) \cdot \cos(2\pi + x)]$.

A. $\mathbf{sin^2 x.}$ Begin by rewriting the trig terms using periodicity identities: $1 - (\sin x \cdot \cot x \cdot \cos x)$. Next, rewrite cotangent in terms of sine and cosine: $1 - (\sin x \cdot {}^{\cos x}\!/_{\sin x} \cdot \cos x)$. Then, cancel the sine from the numerator and denominator, leaving you with $1 - \cos^2 x$. Using Pythagorean identities, this is the same as $\sin^2 x$.

---

17. Simplify $\cos(2\pi + x) + \sin(2\pi + x) \cdot \cot(\pi + x)$.

Solve It

18. Simplify $\dfrac{\cos(x + 4\pi)}{\cot(x + 2\pi)}$.

Solve It

---

19. Simplify $\dfrac{\sec(x + 2\pi)}{\csc(x + 2\pi)}$.

Solve It

20. Simplify $[\sec(x - 2\pi) - \tan(x - \pi)][\sec(x + 2\pi) + \tan(x + \pi)]$.

Solve It

Tackling Trig Proofs

Proofs?!? You thought you left those behind in geometry. Nope, sorry. Don't worry, though — we walk you through pre-calc's version of proofs, which are trig proofs. One thing to remember is that you're just using what you've already practiced in this chapter. These proofs are composed of two sides of an equation. Your job is to make one side look like the other. Here are some hints on how to solve these:

- ✔ **Deal with fractions using basic fraction rules:** The same rules apply to simplifying trig expressions as any other expression. Two key rules to remember:
  - Dividing a fraction by another fraction is the same as multiplying by the reciprocal.
  - Use lowest common denominator (or LCD) when adding or subtracting fractions.
- ✔ **Factor when you can:** Keep an eye out for factorable terms, including factoring out the greatest common factor (GCF) and factoring trinomials (see Chapter 4).
- ✔ **Square square roots:** When you have a square root in a proof, you probably have to square both sides of the proof.
- ✔ **Work on the more complicated side first:** Because the goal is to make one side look like the other, it's generally easier to start on the more complicated side first. If you get stuck, try working on the other side for a while. You can then work backward to simplify the first side.

Be careful with this last hint — some teachers require the proof be done to one side only, so take a look at the other side of the equation, but only to help you get to the end. In *Pre-Calculus For Dummies,* we show you exactly how to do this if you get a teacher who *insists* that you work only on one side.

Q. Prove $\dfrac{1}{\cot^2 x} - \dfrac{1}{\csc^2 x} = \dfrac{\sin^4 x}{\cos^2 x}$.

A. The left side is more complicated, so we'll work on that side. Start by finding the common denominator and add the fractions.

$$\frac{1}{\cot^2 x} - \frac{1}{\csc^2 x} = \frac{\csc^2 x}{\cot^2 x \csc^2 x} - \frac{\cot^2 x}{\cot^2 x \csc^2 x} = \frac{\csc^2 x - \cot^2 x}{\cot^2 x \csc^2 x} = \frac{1}{\cot^2 x \csc^2 x} =$$

From there, notice that you have a Pythagorean identity. Rewrite the resulting fraction in terms of sine and cosine using ratio identities, and multiply terms to complete the proof.

$$\frac{1}{\cot^2 x} \cdot \frac{1}{\csc^2 x} = \tan^2 x \cdot \sin^2 x = \frac{\sin^2 x}{\cos^2 x} \cdot \sin^2 x = \frac{\sin^4 x}{\cos^2 x}$$

21. Prove $\dfrac{\cot x - 1}{1 + \tan(-x)} = \cot x$.

Solve It

22. Prove $\dfrac{\csc x + \tan x}{\cot(\frac{\pi}{2} - x)\csc(\frac{\pi}{2} - x)} = \cot^2 x + \cos x$.

Solve It

23. Prove $\cot x = \dfrac{\csc^2 x - 1}{\cot x}$.

Solve It

24. Prove $\dfrac{1 - \sin x}{\csc x} - \dfrac{\tan x \cos x}{1 + \sin x} = \dfrac{\sin^2 x}{\csc x + 1}$.

Solve It

25. Prove $\sqrt{[\cos x \cdot \sin(\frac{\pi}{2} - x) \cdot \csc(2\pi + x) \cdot \sec(\frac{\pi}{2} - x)] + 1} = \csc x$.

Solve It

26. Prove $\sec x - \cos x = \sin x \cdot \tan x$.

Solve It

Answers to Problems on Basic Trig Identities

1 Simplify $\cot x \cdot \sec x$. **The answer is $\csc x$.**

Start by using ratio identities to show the expression in terms of sine and cosine. Cancel the cosine in the numerator and denominator and write as a single fraction. Finally, rewrite using reciprocal identities.

$$\cot x \cdot \sec x = \frac{\cos x}{\sin x} \cdot \frac{1}{\cos x} = \frac{\cancel{\cos x}}{\sin x} \cdot \frac{1}{\cancel{\cos x}} = \frac{1}{\sin x} = \csc x$$

2 Simplify $\sin x \cdot \sec x$. **The answer is $\tan x$.**

Begin by rewriting $\sec x$ using reciprocal identities. Then multiply the terms to get a single fraction. Finally, rewrite as a single expression using ratio identities.

$$\sin x \cdot \sec x = \sin x \cdot \frac{1}{\cos x} = \frac{\sin x}{\cos x} = \tan x$$

3 Simplify $\sin^3 x \cdot \csc^2 x + \tan x \cdot \cos x$. **The answer is $2 \cdot \sin x$.**

Start by using reciprocal and ratio identities to rewrite the expression in terms of sine and cosine. Next, cancel any terms you can. Finally, combine the like terms.

$$\sin^3 x \cdot \csc^2 x + \tan x \cdot \cos x = \sin^3 x \cdot \frac{1}{\sin^2 x} + \frac{\sin x}{\cos x} \cdot \cos x =$$

$$\sin^{3^1} x \cdot \frac{1}{\cancel{\sin^2 x}} + \frac{\sin x}{\cancel{\cos x}} \cdot \cancel{\cos x} = \sin x + \sin x = 2 \sin x$$

Careful that you write $\sin x + \sin x = 2\sin x$, not $\sin 2x$; this is a double angle (which we introduce in the next chapter).

4 Simplify $\cot x \cdot \sin x \cdot \tan x$. **The answer is $\sin x$.**

Here, you may want to start by rewriting the tangent term using ratio identities. Then, cancel the like terms in the numerator and denominator, giving you your answer.

$$\cot x \cdot \sin x \cdot \tan x = \cot x \cdot \sin x \cdot \frac{1}{\cot x} = \cancel{\cot x} \cdot \sin x \cdot \frac{1}{\cancel{\cot x}} = \sin x$$

5 Simplify $\frac{\cos x}{\sin x}(\tan x + \cot x)$. **The answer is $\csc^2 x$.**

Begin by using ratio identities to rewrite the ratio of cosine and sine as cotangent. Then distribute the cotangent term and rewrite the tangent term using ratio identities, canceling terms in the numerator and denominator. Notice that the resulting expression is a Pythagorean identity that can be simplified.

$$\frac{\cos x}{\sin x}(\tan x + \cot x) = \cot x(\tan x + \cot x) = \cot x \cdot \tan x + \cot x \cdot \cot x =$$

$$\cot x \cdot \frac{1}{\cot x} + \cot^2 x = \cancel{\cot x} \cdot \frac{1}{\cancel{\cot x}} + \cot^2 x = 1 + \cot^2 x = \csc^2 x$$

6 Simplify $1 - \dfrac{\sin x \cdot \tan x}{\sec x}$. **The answer is $\cos^2 x$.**

Again, begin by using reciprocal identities to rewrite tangent and secant in terms of sine and cosine. Next, rewrite the fraction as a division problem, then multiply by the reciprocal. Cancel terms in the numerator and denominator. Finally, use Pythagorean identities to rewrite the expression as a single term.

$$1 - \frac{\sin x \cdot \tan x}{\sec x} = 1 - \frac{\sin x \cdot \dfrac{\sin x}{\cos x}}{\dfrac{1}{\cos x}} = 1 - \sin x \cdot \frac{\sin x}{\cos x} \div \frac{1}{\cos x} =$$

$$1 - \sin x \cdot \frac{\sin x}{\cos x} \cdot \cos x = 1 - \sin x \cdot \frac{\sin x}{\cos x} \cdot \cos x = 1 - \sin^2 x = \cos^2 x$$

7 Simplify $\sin x \cdot \cot^2 x + \sin x$. **The answer is $\csc x$.**

Factor $\sin x$ from both terms. Then, using Pythagorean identities, simplify the expression. Next, rewrite the cosecant term using reciprocal identities. Cancel the like terms in the numerator and denominator. Finally, use reciprocal identities to simplify the final term.

$$\sin x \cdot \cot^2 x + \sin x = \sin x(\cot^2 x + 1) = \sin x \cdot \csc^2 x =$$

$$\sin x \cdot \frac{1}{\sin^2 x} = \sin x \cdot \frac{1}{\sin^2 x} = \frac{1}{\sin x} = \csc x$$

8 Simplify $(\sin^2 x - 1)(\tan^2 x + 1)$. **The answer is 1.**

Start by replacing the sine term using Pythagorean identities. Next, distribute the cosine term across the tangent expression. Then, using ratio identities, rewrite the tangent term. You can then cancel cosine from the numerator and denominator. The resulting expression is another Pythagorean identity.

$$(\sin^2 x - 1)(\tan^2 x + 1) = \cos^2 x(\tan^2 x + 1) = \cos^2 x \cdot \tan^2 x + \cos^2 x =$$

$$\cos^2 x \cdot \frac{\sin^2 x}{\cos^2 x} + \cos^2 x = \cos^2 x \cdot \frac{\sin^2 x}{\cos^2 x} + \cos^2 x = \sin^2 x + \cos^2 x = 1$$

9 Simplify $\sec(-x) \cdot \cot(-x)$. **The answer is $-\csc x$.**

Begin by using even-odd identities to get rid of all the $-x$ values inside the trig expressions. Next, use reciprocal and ratio identities to rewrite the expression in terms of sine and cosine. Then, cancel any like terms in the numerator and denominator. Last, rewrite the resulting fraction using reciprocal identities.

$$\sec(-x) \cdot \cot(-x) = \sec x \cdot (-\cot x) = \frac{1}{\cos x} \cdot \left(-\frac{\cos x}{\sin x} \right) =$$

$$\frac{1}{\cos x} \cdot \left(-\frac{\cos x}{\sin x} \right) = -\frac{1}{\sin x} = -\csc x$$

10 Simplify $\sin x \cdot [\csc x + \sin(-x)]$. **The answer is $\cos^2 x$.**

Use even-odd identities to replace the $-x$ value; then you can distribute the $\sin x$ term. Using reciprocal identities, rewrite the cosecant term and cancel where you can. The resulting expression can be simplified using Pythagorean identities.

$$\sin x \cdot [\csc x + \sin(-x)] = \sin x \cdot [\csc x - \sin x] = \sin x \cdot \csc x - \sin x \cdot \sin x =$$

$$\sin x \cdot \frac{1}{\sin x} - \sin^2 x = \sin x \cdot \frac{1}{\sin x} - \sin^2 x = 1 - \sin^2 x = \cos^2 x$$

11 Simplify $\dfrac{\csc x}{\cot(-x)}$. **The answer is –sec$x$.**

Start by replacing the $-x$ value using even-odd identities. Then, using reciprocal and ratio identities, rewrite everything in terms of sine and cosine. You can then rewrite the big division bar with a division sign and then invert the fraction so you can multiply. Cancel any terms you can. Finally, simplify the resulting expression using reciprocal identities.

$$\frac{\csc x}{\cot(-x)} = \frac{\csc x}{-\cot x} = \frac{\dfrac{1}{\sin x}}{-\dfrac{\cos x}{\sin x}} = \frac{1}{\sin x} \div \left(-\frac{\cos x}{\sin x}\right) =$$

$$\frac{1}{\sin x} \cdot \left(-\frac{\sin x}{\cos x}\right) = \frac{1}{\sin x} \cdot \left(-\frac{\sin x}{\cos x}\right) = -\frac{1}{\cos x} = -\sec x$$

12 Simplify $\dfrac{\cot^2(-x)-1}{\cot^2 x}$. **The answer is sec$^2x$.**

Again, you can start by replacing the $-x$ using even-odd identities. Then you can factor out a negative from the numerator. This results in a Pythagorean identity, which can be simplified. At this point the negatives in the numerator and denominator cancel each other out, leaving you with a positive expression. Next, using reciprocal and ratio identities, change everything into terms using sine and cosine. Rewrite the big fraction as a division problem, which can then be changed to multiplication by inverting the fraction. Cancel what you can and use reciprocal identities to simplify. *Voilà!*

$$\frac{\cot^2(-x)-1}{\cot^2 x} = \frac{-\cot^2 x-1}{-\cot^2 x} = \frac{-(\cot^2 x+1)}{-\cot^2 x} = \frac{\csc^2 x}{\cot^2 x} = \frac{\dfrac{1}{\sin^2 x}}{\dfrac{\cos^2 x}{\sin^2 x}} =$$

$$\frac{1}{\sin^2 x} \div \frac{\cos^2 x}{\sin^2 x} = \frac{1}{\sin^2 x} \cdot \frac{\sin^2 x}{\cos^2 x} = \frac{1}{\sin^2 x} \cdot \frac{\sin^2 x}{\cos^2 x} = \frac{1}{\cos^2 x} = \sec^2 x$$

13 Simplify $\sin\left(\dfrac{\pi}{2} - x\right) \cdot \cot(\pi - x)$. **The answer is sin$x$.**

Here you can start by using co-function identities to rewrite the sine and cotangent terms. Next, use ratio identities to rewrite the tangent term as a rational function of sine and cosine. Cancel the cosine terms, leaving you with your answer.

$$\sin(\pi/2 - x) \cdot \cot(\pi/2 - x) = \cos x \cdot \tan x = \cos x \cdot \frac{\sin x}{\cos x} = \cos x \cdot \frac{\sin x}{\cos x} = \sin x$$

14 Simplify $\sin\left(\dfrac{\pi}{2} - x\right) + \cot\left(\dfrac{\pi}{2} - x\right) \cdot \cos\left(\dfrac{\pi}{2} - x\right)$. **The answer is sec$x$.**

Begin by using co-function identities to replace any terms with $\pi/2$ in them. Then rewrite tangent using ratio identities. Next, you need to find a common denominator in order to add the resulting terms. Rewriting the resulting fraction, you can see that the numerator is a Pythagorean identity, which can be simplified. Finally, use reciprocal identities to simplify the resulting expression.

$$\sin(\pi/2 - x) + \cot(\pi/2 - x) \cdot \cos(\pi/2 - x) = \cos x + \tan x \cdot \sin x = \cos x + \frac{\sin x}{\cos x} \cdot \sin x =$$

$$\cos x + \frac{\sin^2 x}{\cos x} = \frac{\cos^2 x}{\cos x} + \frac{\sin^2 x}{\cos x} = \frac{\cos^2 x + \sin^2 x}{\cos x} = \frac{1}{\cos x} = \sec x$$

15 Simplify $\dfrac{\cos(\pi/2 - x)}{\sin(\pi/2 - x)} \cdot \cot(\pi/2 - x)$. **The answer is $\tan^2 x$.**

Begin by using co-function identities to replace each term. Then use ratio identities to replace the sine/cosine fraction with tangent. Finally, multiply the tangent terms.

$$\frac{\cos(\pi/2 - x)}{\sin(\pi/2 - x)} \cdot \cot(\pi/2 - x) = \frac{\sin x}{\cos x} \cdot \tan x = \tan x \cdot \tan x = \tan^2 x$$

16 Simplify $\left[\sec\left(\dfrac{\pi}{2} - x\right) + \tan\left(\dfrac{\pi}{2} - x\right) \right]\left[1 - \sin\left(\dfrac{\pi}{2} - x\right) \right]$. **The answer is $\sin x$.**

Start by replacing terms using co-function identities. Next, use reciprocal and ratio identities to rewrite the terms using sine and cosine. Then, because the first two terms have a common denominator of sine, you can write them as a single fraction. Next, you can multiply using FOIL. The resulting numerator is a Pythagorean identity which can be simplified. Finally, cancel sine from the numerator and denominator.

$$[\sec(\pi/2 - x) + \tan(\pi/2 - x)][1 - \sin(\pi/2 - x)] = (\csc x + \cot x)(1 - \cos x) =$$

$$\left(\frac{1}{\sin x} + \frac{\cos x}{\sin x} \right)(1 - \cos x) = \left(\frac{1 + \cos x}{\sin x} \right)\left(\frac{1 - \cos x}{1} \right) = \frac{1 - \cos^2 x}{\sin x} = \frac{\sin^2 x}{\sin x} = \frac{\sin^{2} x}{\sin x} = \sin x$$

17 Simplify $\cos(2\pi + x) + \sin(2\pi + x) \cdot \cot(\pi + x)$. **The answer is $2\cos x$.**

Using periodicity identities, replace each term. Then, using ratio identities, replace the cotangent with a ratio of cosine and sine. Cancel the sine from the numerator and denominator and add the two cosines.

$$\cos(2\pi + x) + \sin x(2\pi + x) \cdot \cot(\pi + x) = \cos x + \sin x \cdot \cot x =$$

$$\cos x + \sin x \cdot \frac{\cos x}{\sin x} = \cos x + \sin x \cdot \frac{\cos x}{\sin x} = \cos x + \cos x = 2\cos x$$

18 Simplify $\dfrac{\cos(x + 4\pi)}{\cot(x + 2\pi)}$. **The answer is $\sin x$.**

Ah! Don't let this one trick you. 4π is a multiple of 2π, so adding that to x also gives you a periodicity identity. Replacing the terms with the appropriate periodicity identity, you can easily simplify this problem. The next step is to use ratio identities to replace the cotangent term with a ratio of cosine and sine. Then, simplify the complex fraction by multiplying the numerator by the reciprocal of the denominator. Cancel where you can to get a simplified expression.

$$\frac{\cos(x + 4\pi)}{\cot(x + 2\pi)} = \frac{\cos x}{\cot x} = \frac{\cos x}{\frac{\cos x}{\sin x}} = \cos x \div \frac{\cos x}{\sin x} = \cos x \cdot \frac{\sin x}{\cos x} = \cos x \cdot \frac{\sin x}{\cos x} = \sin x$$

19 Simplify $\dfrac{\sec(x + 2\pi)}{\csc(x + 2\pi)}$. **The answer is $\tan x$.**

Start by replacing both terms using periodicity identities. Then, you can rewrite the fraction as a division problem. Next, rewrite the division problem by multiplying by the reciprocal. Then use reciprocal identities to replace both terms. Multiply to write as a single term, which can be simplified using ratio identities.

$$\frac{\sec(x + 2\pi)}{\csc(x + 2\pi)} = \frac{\sec x}{\csc x} = \sec x \div \csc x = \sec x \cdot \frac{1}{\csc x} = \frac{1}{\cos x} \cdot \sin x = \frac{\sin x}{\cos x} = \tan x$$

20 Simplify $[\sec(x - 2\pi) - \tan(x - \pi)][\sec(x + 2\pi) + \tan(x + \pi)]$. **The answer is 1.**

Simplify by replacing every term using periodicity identities. Then you can FOIL. Finally, you have a Pythagorean identity that can be simplified.

$$\left[\sec(x - 2\pi) - \tan(x - 2\pi) \right]\left[\sec(x + 2\pi) + \tan(x + 2\pi) \right] =$$
$$(\sec x - \tan x)(\sec x + \tan x) = \sec^2 x - \tan^2 x = 1$$

21 Prove $\dfrac{\cot x - 1}{1 + \tan(-x)} = \cot x$.

For this proof, we start with the left side because it's more complicated. Begin by using even-odd identities to replace the $-x$ and use ratio identities to rewrite tangent and cotangent in terms of sine and cosine.

$$\frac{\cot x - 1}{1 + \tan(-x)} = \frac{\cot x - 1}{1 - \tan x} = \frac{\dfrac{\cos x}{\sin x} - 1}{1 - \dfrac{\sin x}{\cos x}}$$

Now multiply the numerator and denominator by the LCD to simplify the complex fraction.

$$\frac{\left(\dfrac{\cos x}{\sin x} - 1\right)\left(\sin x \cdot \cos x\right)}{\left(1 - \dfrac{\sin x}{\cos x}\right)\left(\sin x \cdot \cos x\right)} = \frac{\dfrac{\cos x}{\sin x}\sin x \cdot \cos x - \sin x \cdot \cos x}{\sin x \cdot \cos x - \dfrac{\sin x}{\cos x}\sin x \cdot \cos x} =$$

$$\frac{\dfrac{\cos x}{\sin x}\sin x \cdot \cos x - \sin x \cdot \cos x}{\sin x \cdot \cos x - \dfrac{\sin x}{\cos x}\sin x \cdot \cos x} = \frac{\cos^2 x - \sin x \cdot \cos x}{\sin x \cdot \cos x - \sin^2 x}$$

Finally, you can pull out a common factor on both the numerator and denominator, cancel any like terms, and simplify the resulting fraction using ratio identities.

$$\frac{\cos^2 x - \sin x \cdot \cos x}{\sin x \cdot \cos x - \sin^2 x} = \frac{\cos x\left(\cos x - \sin x\right)}{\sin x\left(\cos x - \sin x\right)} = \frac{\cos x\left(\cos x - \sin x\right)}{\sin x\left(\cos x - \sin x\right)} = \frac{\cos x}{\sin x} = \cot x$$

22 Prove $\dfrac{\csc x + \tan x}{\cot(\sfrac{\pi}{2} - x)\csc(\sfrac{\pi}{2} - x)} = \cot^2 x + \cos x$.

Start by using co-function identities to replace the terms with π.

$$\frac{\csc x + \tan x}{\cot\left(\sfrac{\pi}{2} - x\right)\csc\left(\sfrac{\pi}{2} - x\right)} = \frac{\csc x + \tan x}{\tan x \sec x} =$$

Next, you can separate the fraction into two different fractions and then cancel any terms on both the numerator and denominator.

$$\frac{\csc x}{\tan x \sec x} + \frac{\tan x}{\tan x \sec x} = \frac{\csc x}{\tan x \sec x} + \frac{\tan x}{\tan x \sec x} = \frac{\csc x}{\tan x \sec x} + \frac{1}{\sec x} =$$

Then, use reciprocal and ratio identities to rewrite all terms using sine and cosine. You can simplify the denominator of the complex fraction using multiplication.

$$\frac{\dfrac{1}{\sin x}}{\dfrac{\sin x}{\cos x} \cdot \dfrac{1}{\cos x}} + \cos x = \frac{\dfrac{1}{\sin x}}{\dfrac{\sin x}{\cos^2 x}} + \cos x =$$

Next, get rid of the complex fraction by changing the large fraction bar to division and then multiplying by the reciprocal. The resulting fraction can also be simplified using ratio identities, giving you the answer.

$$\frac{1}{\sin x} \div \frac{\sin x}{\cos^2 x} + \cos x = \frac{1}{\sin x} \cdot \frac{\cos^2 x}{\sin x} + \cos x = \frac{\cos^2 x}{\sin^2 x} + \cos x = \cot^2 x + \cos x$$

23 Prove $\cot x = \dfrac{\csc^2 x - 1}{\cot x}$.

In this proof, the right side is more complicated, so it's wise to start there. Notice the squared term? You can replace it using Pythagorean identities, which then cancels out the 1 in the numerator. Last, simply cancel a cotangent from both the numerator and denominator and you're there!

$$\cot x = \frac{\csc^2 x - 1}{\cot x} = \frac{\left(\cot^2 x + 1\right) - 1}{\cot x} = \frac{\cot^2 x}{\cot x} = \frac{\cot^2 x}{\cot x} = \cot x$$

24 Prove $\dfrac{1 - \sin x}{\csc x} - \dfrac{\tan x \cos x}{1 + \sin x} = \dfrac{\sin^2 x}{\csc x + 1}$.

For this proof, you want to start by finding a common denominator for the two fractions and multiplying it through using FOIL for the first fraction. Next, you can simplify using a Pythagorean identity.

$$\frac{1 - \sin x}{\csc x} - \frac{\tan x \cos x}{1 + \sin x} = \frac{\left(1 - \sin x\right)\left(1 + \sin x\right)}{\csc x\left(1 + \sin x\right)} - \frac{\tan x \cos x \csc x}{\left(1 + \sin x\right)\csc x} =$$

$$\frac{1 - \sin^2 x}{\csc x\left(1 + \sin x\right)} - \frac{\tan x \cos x \csc x}{\left(1 + \sin x\right)\csc x} = \frac{\cos^2 x}{\csc x\left(1 + \sin x\right)} - \frac{\tan x \cos x \csc x}{\left(1 + \sin x\right)\csc x} =$$

After you have a single fraction, use reciprocal and ratio identities to change the numerator, allowing you to cancel many of the terms. Again, you have a Pythagorean identity that you can simplify.

$$\frac{\cos^2 x - \dfrac{\sin x}{\cos x} \cdot \cos x \cdot \dfrac{1}{\sin x}}{\csc x\left(1 + \sin x\right)} = \frac{\cos^2 x - \dfrac{\sin x}{\cos x} \cdot \cos x \cdot \dfrac{1}{\sin x}}{\csc x\left(1 + \sin x\right)} =$$

$$\frac{\cos^2 x - 1}{\csc x\left(1 + \sin x\right)} = \frac{\sin^2 x}{\csc x\left(1 + \sin x\right)} =$$

Now that the numerator looks like your final answer, you can concentrate on the denominator. Distribute the cosecant and use reciprocal identities to simplify.

$$\frac{\sin^2 x}{\csc x + \csc x \sin x} = \frac{\sin^2 x}{\csc x + \dfrac{1}{\sin x} \cdot \sin x} = \frac{\sin^2 x}{\csc x + \dfrac{1}{\sin x} \cdot \sin x} = \frac{\sin^2 x}{\csc x + 1}$$

25 Prove $\sqrt{[\cos x \cdot \sin(\pi/2 - x) \cdot \csc(2\pi + x) \cdot \sec(\pi/2 - x)] + 1} = \csc x$.

Start by dealing with the gigantic square root by squaring both sides.

$$\sqrt{[\cos x \cdot \sin(\pi/2 - x) \cdot \csc(2\pi + x) \cdot \sec(\pi/2 - x)] + 1} = \csc x$$

$$[\cos x \cdot \sin(\pi/2 - x) \cdot \csc(2\pi + x) \cdot \sec(\pi/2 - x)] + 1 = \csc^2 x$$

Next, replace any terms with $\pi/2$ using co-function identities, and replace any terms with 2π using periodicity identities. Multiply the resulting trig terms.

$$\cos x \cdot \cos x \cdot \csc x \cdot \csc x + 1 =$$

$$\cos^2 x \cdot \csc^2 x + 1 =$$

Replace the cosecant term using reciprocal identities and write as a single fraction that you can replace using ratio identities.

$$\cos^2 x \cdot \frac{1}{\sin^2 x} + 1 = \frac{\cos^2 x}{\sin^2 x} + 1 = \cot^2 x + 1 =$$

Finally, you have a Pythagorean identity that can be simplified to get the answer you want.

$$\csc^2 x$$

26 Prove $\sec x - \cos x = \sin x \cdot \tan x$.

Starting by replacing secant with its reciprocal identity, you end up with a fraction. Adding the fractions using a common denominator of cosine, you get a Pythagorean identity in the numerator, which can be simplified. Finally, multiplying out a sine leaves you with a ratio of sine and cosine that you can replace using ratio identities to get the right side of the proof.

$$\sec x - \cos x = \frac{1}{\cos x} - \cos x = \frac{1}{\cos x} - \frac{\cos^2 x}{\cos x} = \frac{1 - \cos^2 x}{\cos x} =$$

$$\frac{\sin^2 x}{\cos x} = \sin x \cdot \frac{\sin x}{\cos x} = \sin x \tan x$$

Chapter 9
Advanced Identities

In This Chapter
▶ Using sum and difference identities
▶ Simplifying with double angle identities
▶ Discovering half-angle identities
▶ Tapping product to sum and sum to product identities
▶ Solving with power-reducing formulas

Okay . . . the training wheels are off! We're getting into the advanced stuff here — advanced identities, that is. This chapter builds on the basic identities you practiced in Chapter 8.

In this chapter, we give you formulas that are essential for calculating precise values of angles that you could never get before (even your calculator only gives approximate answers). These identities are essential for calculus and are well-loved by pre-calc teachers, so it's time to get friendly with advanced identities.

Simplifying with Sum and Difference Identities

Here those pesky mathematicians took simple concepts (addition and subtraction) and related them to a more complex one (trigonometric angles). These *sum and difference identities* allow you to write an angle that's not from the special triangles of 45-45-90 or 30-60-90 (see Chapter 6) as the sum or difference of those helpful angle measures. For example, you can rewrite the measure of 105° as the sum of 45° and 60°. The problems presented here (and in pre-calc books everywhere) can always be written using the angles you already have exact values for, even though these identities can be used for any value.

The sum and difference identities are:

$$\sin(a \pm b) = \sin a \cdot \cos b \pm \cos a \cdot \sin b$$
$$\cos(a \pm b) = \cos a \cdot \cos b \mp \sin a \cdot \sin b$$
$$\tan(a \pm b) = \frac{\tan a \pm \tan b}{1 \mp \tan a \cdot \tan b}$$

Q. Find $\tan \frac{7\pi}{12}$ using sum or difference identities.

A. $-2 - \sqrt{3}$. Start by breaking up the fraction into a sum of two values that can be found on the unit circle: $\frac{7\pi}{12} = \frac{3\pi}{12} + \frac{4\pi}{12} = \frac{\pi}{4} + \frac{\pi}{3}$. Next, plug the angles into the sum identity for

tangent: $\tan\left(\dfrac{\pi}{4} + \dfrac{\pi}{3}\right) = \dfrac{\tan \dfrac{\pi}{4} + \tan \dfrac{\pi}{3}}{1 - \tan \dfrac{\pi}{4} \cdot \tan \dfrac{\pi}{3}}$.

Keep a close eye on the order of a plus or minus symbol in an equation. If it's inverted to be a minus or plus symbol, then you perform the opposite operation than the given problem. In this case, the problem involves addition, so we use addition in the numerator and subtraction in the denominator.

Finally, plug in the known values for the angles from the unit circle (see the Cheat Sheet) and simplify. Remember to rationalize the denominator using conjugates (see Chapter 2 for review). The steps are as follows:

$$\dfrac{\tan \dfrac{\pi}{4} + \tan \dfrac{\pi}{3}}{1 - \tan \dfrac{\pi}{4} \cdot \tan \dfrac{\pi}{3}} =$$

$$\dfrac{1 + \sqrt{3}}{1 - (1)\left(\sqrt{3}\right)} =$$

$$\dfrac{1 + \sqrt{3}}{1 - \sqrt{3}} =$$

$$\left(\dfrac{1 + \sqrt{3}}{1 - \sqrt{3}}\right)\left(\dfrac{1 + \sqrt{3}}{1 + \sqrt{3}}\right) =$$

$$\dfrac{1 + 2\sqrt{3} + 3}{1 - 3} =$$

$$\dfrac{4 + 2\sqrt{3}}{-2} =$$

$$-2 - \sqrt{3}$$

---

1. Find $\cos 15°$ using sum or difference identities.

Solve It

2. Simplify $\tan(45° - x)$.

Solve It

---

3. Prove $\dfrac{\sin\left(x + \pi\right)}{\cos\left(x + \pi\right)} = \tan x$.

Solve It

4. Prove $\dfrac{\sin\left(x + y\right)}{\sin x \cdot \cos y} = 1 + \cot x \cdot \tan y$.

Solve It

5. Find $\csc \frac{5\pi}{12}$ using sum or difference identities.

Solve It

6. Simplify $\sec(180° + x)$ using sum or difference identities.

Solve It

Using Double Angle Identities

Double angle identities help you find the trig value of twice an angle. These can be used to find an exact value if you know the original angle. They can also be used to prove trig proofs (see Chapter 8) or solve trig equations. Cosine has three double angle identities created from the Pythagorean identities of Chapter 8. You have a choice as to which you want to use depending on the problem.

The double angle identities are:

$$\sin 2x = 2\sin x \cos x$$
$$\cos 2x = \cos^2 x - \sin^2 x = 2\cos^2 x - 1 = 1 - 2\sin^2 x$$
$$\tan 2x = \frac{2\tan x}{1 - \tan^2 x}$$

Q. Solve $6\cos^2 x - 6\sin^2 x = 3$ for $\pi > x > 0$.

A. $x = \frac{\pi}{6}$. Start by factoring out the 6 from the left side of the equation: $6(\cos^2 x - \sin^2 x) = 3$. Next, substitute using the appropriate double angle identity: $6(\cos 2x) = 3$. Isolate the trigonometric term: $\cos 2x = \frac{1}{2}$. Then,

take the inverse of cosine using the unit circle (see the Cheat Sheet): $2x = \frac{\pi}{3}$. Finally, solve for x by dividing both sides by 2: $x = \frac{\pi}{6}$. Because you're limited to quadrants I and II and cosine is only positive in quadrants I and IV, there's only the one answer.

7. Find the value of $\cos 2x$ if $\csc x = \frac{13}{5}$.

Solve It

8. Find the value of $\tan 2x$ if $\cot x = \frac{1}{2}$.

Solve It

9. Prove $\dfrac{\cos 2x + 1}{\sin 2x} = \cot x$.

Solve It

10. Prove $\dfrac{\cos 2x}{\sin 2x} = \dfrac{1}{2}\left(\cot x - \tan x\right)$.

Solve It

11. Simplify $\tan 3x$ using double angle identities.

Solve It

12. Solve $6 - 12\sin^2 x = 3\sqrt{3}$ for $\pi > x > 0$.

Solve It

Reducing with Half-Angle Identities

Similar to sum and difference identities, *half-angle identities* help you find exact values of unusual angles, namely ones that are half the value of ones you already know. For example, if we want to find a trig value of 22.5°, we would use the half-angle identity of half of 45° because 22.5 is half of 45. Also, just like every other identity we've reviewed so far, half-angle identities can be used for proving trig proofs and solving trig equations.

The half-angle identities are:

$$\sin\left(\frac{\alpha}{2}\right) = \pm\sqrt{\frac{1-\cos\alpha}{2}}$$

$$\cos\left(\frac{\alpha}{2}\right) = \pm\sqrt{\frac{1+\cos\alpha}{2}}$$

$$\tan\left(\frac{\alpha}{2}\right) = \frac{1-\cos\alpha}{\sin\alpha} = \frac{\sin\alpha}{1+\cos\alpha}$$

Q. Find $\cot \frac{5\pi}{12}$ using half-angle identities.

A. $2-\sqrt{3}$. First, because there isn't a half-angle formula for cotangent, we have to start by recognizing that $\cot \frac{5\pi}{12}$ is the reciprocal of tangent of the same angle. Therefore, we're going to find the value of $\tan \frac{5\pi}{12}$, then take the reciprocal. The angle $\frac{5\pi}{12}$ can be rewritten as $\frac{\frac{5\pi}{6}}{2}$. Plugging that into the half-angle identity, we get $\dfrac{1-\cos\dfrac{5\pi}{6}}{\sin\dfrac{5\pi}{6}}$.

Replacing the trig expressions for the exact values from the unit circle, we get $\dfrac{1-\left(-\dfrac{\sqrt{3}}{2}\right)}{\dfrac{1}{2}}$. This simplifies to $\sqrt{3}$. But wait — we're not done! This is the value of $\tan \frac{5\pi}{12}$, and we need cot $\sqrt{3}$. So we have to take the reciprocal, which, after rationalizing, we find to be $2-\sqrt{3}$. Whew!

---

13. Find $\tan \frac{3\pi}{8}$ using half-angle identities.

Solve It

14. Find $\sin \frac{7\pi}{12}$.

Solve It

15. Prove $\left[\tan\left(\dfrac{x}{2}\right)\right]\cdot\tan x = \dfrac{\sin^2 x}{\cos x + \cos^2 x}$.

Solve It

16. Find an approximate value of $\cos\frac{5}{24}$ using half-angle identities.

Solve It

Changing Products to Sums

You know that age-old question about math — when will I ever use this stuff? Well, this is one section in pre-calc that even stumps us, because the only time you'll ever need *product to sum (or difference) identities* is in calculus; they don't really have any real-world uses. So, do you still need to learn about them? Yep, 'fraid so — pre-calc and calc teachers expect you to know them, so here we go.

You need to know three product to sum identities: sin · cos, cos · cos, and sin · sin. Here they are:

$$\sin a \cdot \cos b = \frac{1}{2}\left[\sin\left(a+b\right)+\sin\left(a-b\right)\right]$$

$$\cos a \cdot \cos b = \frac{1}{2}\left[\cos\left(a+b\right)+\cos\left(a-b\right)\right]$$

$$\sin a \cdot \sin b = \frac{1}{2}\left[\cos\left(a-b\right)-\cos\left(a+b\right)\right]$$

Q. Express $8\sin 3x \cdot \sin x$ as a sum or difference.

A. **4(cos2x – cos4x).** Start by plugging in the appropriate product to sum identity: $8\left(\frac{1}{2}\left[\cos\left(3x-x\right)-\cos\left(3x+x\right)\right]\right)$. Then, simplify using multiplication: $8\cdot\frac{1}{2}\left[\cos\left(2x\right)-\cos\left(4x\right)\right]=4\left(\cos 2x-\cos 4x\right)$.

17. Express $12\cos6x \cdot \sin2x$ as a sum or difference.

Solve It

18. Express $2\sin5x \cdot \cos2x$ as a sum or difference.

Solve It

19. Express $6\cos3x \cdot \sin6x$ as a sum or difference.

Solve It

20. Express $7\sin8x \cdot \sin3x$ as a sum or difference.

Solve It

Expressing Sums as Products

Although less frequently used than the other identities in this chapter, the *sum to product identities* are useful for finding exact answers for some trig expressions. In cases where the sum or difference of the two angles results in an angle from our special right triangles (Chapter 6), then sum to product identities can be quite helpful. The sum (or difference) to product identities involve the addition or subtraction of either sine or cosine.

The sum (or difference) to product identities are:

$$\sin x + \sin y = 2\sin\left(\frac{x+y}{2}\right)\cos\left(\frac{x-y}{2}\right)$$

$$\sin x - \sin y = 2\cos\left(\frac{x+y}{2}\right)\sin\left(\frac{x-y}{2}\right)$$

$$\cos x + \cos y = 2\cos\left(\frac{x+y}{2}\right)\cos\left(\frac{x-y}{2}\right)$$

$$\cos x - \cos y = -2\sin\left(\frac{x+y}{2}\right)\sin\left(\frac{x-y}{2}\right)$$

0. Find $\cos 165° + \cos 75°$.

A. $-\frac{\sqrt{2}}{2}$. Begin by using the sum to product identity to rewrite the expression:

$2\cos\left(\frac{165+75}{2}\right)\cos\left(\frac{165-75}{2}\right)$. Simplify the results using unit circle values:
$2\cos\left(\frac{240}{2}\right) \cdot \cos\left(\frac{90}{2}\right) = 2\cos 120° \cdot \cos 45° =$

$2(-\tfrac{1}{2}) \cdot \frac{\sqrt{2}}{2} = -\frac{\sqrt{2}}{2}$.

21. Find $\sin 195° - \sin 75°$.

Solve It

22. Find $\cos 375° - \cos 75°$.

Solve It

23. Find $\sin \frac{7\pi}{12} + \sin \frac{5\pi}{12}$.

Solve It

24. Find $\cos \frac{23\pi}{12} + \cos \frac{5\pi}{12}$.

Solve It

Powering Down: Power-Reducing Formulas

Power-reducing formulas can be used to simplify trig expressions with exponents and can be used more than once if you have a function that's raised to the fourth power or higher. These nifty formulas help you get rid of exponents so you can focus on solving for an angle's measure.

Here are the three power-reducing formulas:

$$\sin^2 u = \frac{1 - \cos 2u}{2}$$

$$\cos^2 u = \frac{1 + \cos 2u}{2}$$

$$\tan^2 u = \frac{1 - \cos 2u}{1 + \cos 2u}$$

Q. Simplify $\cos^4 x$ using power-reducing formulas.

A. $\frac{1}{8}(3 + 4\cos 2x + \cos 4x)$. After rewriting $(\cos^2 x)^2$, you can see that you need to use the power-reducing formula twice. The first time gives you

$$\left(\frac{1 + \cos 2x}{2}\right)^2 = \frac{(1 + \cos 2x)(1 + \cos 2x)}{4} = \frac{1 + 2\cos 2x + \cos^2 x}{4} = \frac{1}{4}\left(1 + 2\cos 2x + \cos^2 x\right)$$

Use the formula again on the remaining squared term and reduce:

$$\frac{1}{4}\left[1 + 2\cos 2x + \left(\frac{1 + \cos 4x}{2}\right)\right] = \frac{1}{8}\left(2 + 4\cos 2x + 1 + \cos 4x\right) = \frac{1}{8}\left(3 + 4\cos 2x + \cos 4x\right)$$

---

25. Prove $\cos^2 3x - \sin^2 3x = \cos 6x$.

Solve It

26. Prove $(1 + \cos 2x) \cdot \tan^3 x = \tan x(1 - \cos 2x)$.

Solve It

---

27. Express $\sin^4 x - \cos^4 x$ without exponents.

Solve It

28. Express $\cot^2(\frac{x}{2})$ without exponents.

Solve It

Answers to Problems on Advanced Identities

1 Find $\cos 15°$ using sum or difference identities. **The answer is $\dfrac{\sqrt{6} - \sqrt{2}}{4}$.**

Start by rewriting the angle using special angles from the unit circle. We chose $15° = 45° - 30°$. Plugging these values into the difference formula, you get $\cos(45° - 30°) = \cos 45° \cdot \cos 30° - \sin 45° \cdot \sin 30°$. Use the unit circle to plug in the appropriate values and simplify:

$$\frac{\sqrt{2}}{2} \cdot \frac{\sqrt{3}}{2} - \frac{\sqrt{2}}{2} \cdot \frac{1}{2} = \frac{\sqrt{6}}{4} - \frac{\sqrt{2}}{4} = \frac{\sqrt{6} - \sqrt{2}}{4}.$$

2 Simplify $\tan(45° - x)$. **The answer is $\dfrac{1 - \tan x}{1 + \tan x}$.**

Rewrite the expression using the difference formula: $\dfrac{\tan 45 - \tan x}{1 + \tan 45 \tan x}$. Then, replace $\tan 45°$ with its unit circle value of 1 and simplify the expression: $\dfrac{(1) - \tan x}{1 + (1)\tan x} = \dfrac{1 - \tan x}{1 + \tan x}$.

3 Prove $\dfrac{\sin(x + \pi)}{\cos(x + \pi)} = \tan x$.

Begin the proof by rewriting the left side using sum formulas. Next, substitute for unit circle values and simplify. Finally, replace the sine and cosine with tangent using ratio identities (from Chapter 8). The steps are as follows:

$$\frac{\sin(x + \pi)}{\cos(x + \pi)} = \tan x$$

$$\frac{\sin x \cdot \cos \pi + \cos x \cdot \sin \pi}{\cos x \cdot \cos \pi - \sin x \cdot \sin \pi} = \tan x$$

$$\frac{\sin x \cdot (-1) + \cos x \cdot (0)}{\cos x \cdot (-1) - \sin x \cdot (0)} = \tan x$$

$$\frac{-\sin x + 0}{-\cos x + 0} = \tan x$$

$$\frac{-\sin x}{-\cos x} = \tan x$$

$$\frac{\sin x}{\cos x} = \tan x$$

$$\tan x = \tan x$$

4 Prove $\dfrac{\sin(x + y)}{\sin x \cdot \cos y} = 1 + \cot x \cdot \tan y$.

Start by replacing the $\sin(x + y)$ term using sum formula. Next, separate the fraction into the sum of two fractions and reduce terms. Last, rewrite using ratio identities from Chapter 8.

$$\frac{\sin(x+y)}{\sin x \cos y} = 1 + \cot x \cdot \tan y$$

$$\frac{\sin x \cdot \cos y + \cos x \cdot \sin y}{\sin x \cos y} = 1 + \cot x \cdot \tan y$$

$$\frac{\sin x \cdot \cos y}{\sin x \cos y} + \frac{\cos x \cdot \sin y}{\sin x \cos y} = 1 + \cot x \cdot \tan y$$

$$\frac{\cancel{\sin x} \cdot \cancel{\cos y}}{\cancel{\sin x} \cancel{\cos y}} + \frac{\cos x \cdot \sin y}{\sin x \cos y} = 1 + \cot x \cdot \tan y$$

$$1 + \frac{\cos x}{\sin x} \cdot \frac{\sin y}{\cos y} = 1 + \cot x \cdot \tan y$$

$$1 + \cot x \cdot \tan y = 1 + \cot x \cdot \tan y$$

5 Find $\csc^{5\pi}\!/_{12}$ using sum or difference identities. **The answer is** $\sqrt{6} - \sqrt{2}$.

First, know that you can rewrite $^{5\pi}\!/_{12}$ as $^{2\pi}\!/_{12} + ^{3\pi}\!/_{12}$, which is the same as $^\pi\!/_6 + ^\pi\!/_4$. Substitute this back into the problem: $\csc(^\pi\!/_6 + ^\pi\!/_4)$. Because you don't have a sum identity for cosecant, you need to

find $\dfrac{1}{\sin\left(\dfrac{\pi}{6} + \dfrac{\pi}{4}\right)}$. To ease this calculation, you start by finding $\sin(^\pi\!/_6 + ^\pi\!/_4)$, then invert it. Using

sum identities, $\sin(^\pi\!/_6 + ^\pi\!/_4) = \sin^\pi\!/_6 \cdot \cos^\pi\!/_4 + \cos^\pi\!/_6 \cdot \sin^\pi\!/_4$. Replace the trig expressions with the

appropriate values from the unit circle and simplify: $\dfrac{1}{2} \cdot \dfrac{\sqrt{2}}{2} + \dfrac{\sqrt{3}}{2} \cdot \dfrac{\sqrt{2}}{2} = \dfrac{\sqrt{2}}{4} + \dfrac{\sqrt{6}}{4} = \dfrac{\sqrt{2} + \sqrt{6}}{4}$.

The last step is to invert this solution and simplify by rationalizing the denominator:

$$\frac{4}{\sqrt{2} + \sqrt{6}} = \left(\frac{4}{\sqrt{2} + \sqrt{6}}\right) \cdot \frac{\left(\sqrt{2} - \sqrt{6}\right)}{\left(\sqrt{2} - \sqrt{6}\right)} = \frac{4\left(\sqrt{2} - \sqrt{6}\right)}{2 - 6} = \frac{4\left(\sqrt{2} - \sqrt{6}\right)}{-4} = -\left(\sqrt{2} - \sqrt{6}\right) = \sqrt{6} - \sqrt{2}.$$

6 Simplify $\sec(180° + x)$ using sum or difference identities. **The answer is sec*x*.**

Because you don't have a sum identity for secant, you need to use the one for cosine and then find the reciprocal. Using this identity, you get $\cos 180 \cdot \cos x - \sin 180 \cdot \sin x$. Plugging in values from the unit circle and simplifying, you get $(1) \cdot \cos x - (0) \cdot \sin x = \cos x$. The reciprocal of this is $^1\!/_{\cos x}$, which by reciprocal identities (from Chapter 8) is $\sec x$.

7 Find the value of $\cos 2x$ if $\csc x = ^{12}\!/_5$. **The answer is** $^{47}\!/_{72}$.

Because $\cos 2x = 1 - 2\sin^2 x$, you need to know $\sin x$ to plug it in. No problem! You have the reciprocal: $\csc x$. Therefore, $\sin x = ^5\!/_{12}$. Plugging this in, you get $\cos 2x = 1 - 2(^5\!/_{12})^2 = 1 - 2(^{25}\!/_{144}) = 1 - ^{25}\!/_{72} = ^{47}\!/_{72}$.

8 Find the value of $\tan 2x$ if $\cot x = \frac{1}{2}$. **The answer is** $-^4\!/_3$.

Begin with finding $\tan x$ by taking the reciprocal of $\cot x$: $^2\!/_1 = 2$. Then plug it into the formula for

$\tan 2x$: $\dfrac{2\tan x}{1 - \tan^2 x} = \dfrac{2(2)}{1 - (2)^2} = \dfrac{4}{1 - 4} = -^4\!/_3$.

9 Prove $\dfrac{\cos 2x + 1}{\sin 2x} = \cot x$.

Working with the left side, change all the double angles using the double angle identities. After combining like terms, cancel the 2 and cosine from the numerator and denominator. Finally, rewrite the result using ratio identities (Chapter 8). The steps are as follows:

$$\frac{\cos 2x + 1}{\sin 2x} = \cot x; \quad \frac{2\cos^2 x - 1 + 1}{2\sin x \cos x} = \cot x; \quad \frac{2\cos^2 x}{2\sin x \cos x} = \cot x;$$

$$\frac{\cancel{2}\cos^2 x}{\cancel{2}\sin x \cancel{\cos x}} = \cot x; \quad \frac{\cos x}{\sin x} = \cot x; \quad \cot x = \cot x$$

10 Prove $\dfrac{\cos 2x}{\sin 2x} = \dfrac{1}{2}\left(\cot x - \tan x\right)$.

Begin by using the double angle identities for the left side of the proof: $\dfrac{\cos 2x}{\sin 2x} = \dfrac{\cos^2 x - \sin^2 x}{2\sin x \cos x}$.

Next, separate the single fraction into the difference of two fractions: $\dfrac{\cos^2 x}{2\sin x \cos x} - \dfrac{\sin^2 x}{2\sin x \cos x}$.

Then, cancel any terms in the numerator and denominator:

$\dfrac{\cos^2 x}{2\sin x \cancel{\cos x}} - \dfrac{\sin^2 x}{2\cancel{\sin x} \cos x} = \dfrac{\cos x}{2\sin x} - \dfrac{\sin x}{2\cos x}$. Finally, use ratio identities to replace the fractions

with cotangent and tangent and factor out the GCF: $\dfrac{1}{2}\cot x - \dfrac{1}{2}\tan x = \dfrac{1}{2}\left(\cot x - \tan x\right)$.

11 Simplify $\tan 3x$ using double angle identities. **The answer is** $\dfrac{3\tan x - \tan^3 x}{1 - 3\tan^2 x}$.

Start by separating out the angle into a single and double angle: $\tan(x + 2x)$. Next, use the sum

identity for tangent: $\dfrac{\tan x + \tan 2x}{1 - \tan x \tan 2x}$. Next, replace the double angle terms using the double-angle

formula: $\dfrac{\tan x + \dfrac{2\tan x}{1 - \tan^2 x}}{1 - \tan x \left(\dfrac{2\tan x}{1 - \tan^2 x}\right)}$. Then, to simplify the complex fraction, multiply by

the common denominator: $\dfrac{\tan x + \dfrac{2\tan x}{1 - \tan^2 x}}{1 - \tan x \left(\dfrac{2\tan x}{1 - \tan^2 x}\right)} \cdot \dfrac{1 - \tan^2 x}{1 - \tan^2 x}$. Multiply through:

$$\frac{\tan x\left(1 - \tan^2 x\right) + \dfrac{2\tan x}{1 - \tan^2 x}\left(1 - \tan^2 x\right)}{1\left(1 - \tan^2 x\right) - \tan x\left(\dfrac{2\tan x}{1 - \tan^2 x}\right)\left(1 - \tan^2 x\right)} = \frac{\tan x - \tan^3 x + 2\tan x}{1 - \tan^2 x - 2\tan^2 x}$$

and combine like terms: $\dfrac{3\tan x - \tan^3 x}{1 - 3\tan^2 x}$.

12 Solve $6 - 12\sin^2 x = 3\sqrt{3}$ for $\pi > x > 0$. **The answer is 15°.**

Begin by factoring out 6 from both terms: $6(1 - 2\sin^2 x) = 3\sqrt{3}$. Next, use double angle identities

to replace the trigonometric term: $6(\cos 2x) = 3\sqrt{3}$. Isolate the trig term by dividing both sides by 6 and simplifying: $\cos 2x = \dfrac{3\sqrt{3}}{6} = \dfrac{\cancel{3}\sqrt{3}}{\cancel{6}_2} = \dfrac{\sqrt{3}}{2}$. Then, to solve for x, take the inverse cosine of each side: $2x = \cos^{-1}\left(\dfrac{\sqrt{3}}{2}\right) = 30°$ and divide by 2: $15°$. Because you're limited to quadrants I and II and the answer is positive, there's only one answer: $15°$.

13 Find $\tan{}^{3\!\!/_8}$ using half-angle identities. **The answer is $\sqrt{2}+1$.**

Start by rewriting the angle to find the half angle: $\tan\dfrac{\frac{3\pi}{4}}{2}$. Then, plug it into the half-angle

identity: $\dfrac{1-\cos\frac{3\pi}{4}}{\sin\frac{3\pi}{4}}$; knowing that the angle is in quadrant II tells you that the result for tangent

is positive. Find the exact values for the trigonometric terms on the unit circle: $\dfrac{1-\left(-\frac{\sqrt{2}}{2}\right)}{\frac{\sqrt{2}}{2}}$.

Then simplify the complex fraction:

$\dfrac{1+\frac{\sqrt{2}}{2}}{\frac{\sqrt{2}}{2}} = \dfrac{2+\sqrt{2}}{2} \div \dfrac{\sqrt{2}}{2} = \dfrac{2+\sqrt{2}}{\cancel{2}} \cdot \dfrac{\cancel{2}}{\sqrt{2}} = \dfrac{2+\sqrt{2}}{\sqrt{2}}\left(\dfrac{\sqrt{2}}{\sqrt{2}}\right) = \dfrac{2\sqrt{2}+2}{2} = \sqrt{2}+1$.

14 Find $\sin{}^{7\!\!/_{12}}$. **The answer is $\dfrac{\sqrt{2+\sqrt{3}}}{4}$.**

Begin by rewriting the angle: $\sin\dfrac{\frac{7\pi}{6}}{2}$. Plug the angle into the appropriate half-angle identity:

$\sqrt{\dfrac{1-\cos\frac{7\pi}{6}}{2}}$. Because the angle is in quadrant I, sine will be positive. The next step is to replace

the trig terms with the exact values from the unit circle: $\sqrt{\dfrac{1-\left(\frac{\sqrt{3}}{2}\right)}{2}}$. Simplify the complex fraction:

$\sqrt{\dfrac{1+\frac{\sqrt{3}}{2}}{2}} = \sqrt{\dfrac{\frac{2+\sqrt{3}}{2}}{2}} = \sqrt{\dfrac{2+\sqrt{3}}{2}\cdot\dfrac{1}{2}} = \sqrt{\dfrac{2+\sqrt{3}}{4}}$

15 Prove $\left[\tan\left(\dfrac{x}{2}\right)\right]\cdot\tan x = \dfrac{\sin^2 x}{\cos x + \cos^2 x}$.

Start with the left side by replacing the half-angle term using the appropriate identity:

$\left(\dfrac{\sin x}{1+\cos x}\right)\tan x = \dfrac{\sin^2 x}{\cos x + \cos^2 x}$. Next, use ratio identities to change everything to sine

and cosine: $\left(\dfrac{\sin x}{1+\cos x}\right)\dfrac{\sin x}{\cos x} = \dfrac{\sin^2 x}{\cos x + \cos^2 x}$. Multiply through to complete the proof:

$\dfrac{\sin^2 x}{\cos x + \cos^2 x} = \dfrac{\sin^2 x}{\cos x + \cos^2 x}$.

16 Find an approximate value of $\cos\frac{7}{24}$ using half-angle identities. **The answer is ≈ 0.99.**

Rewriting the angle using half-angle identities, remember that the angle is in quadrant I,

so it's positive: $\cos\dfrac{\frac{\pi}{12}}{2}$. Replace it with the appropriate half-angle identity: $\sqrt{\dfrac{1+\cos\frac{\pi}{12}}{2}}$. Because

you don't have a special right triangle value, you need to use the half-angle identities again:

 . Now you can replace the cosine term using the unit circle value (finally!):

$\sqrt{\dfrac{1+\sqrt{\dfrac{1+\frac{\sqrt{3}}{2}}{2}}}{2}}$. Finish by plugging the complex fraction into your calculator: ≈ 0.99.

17 Express $12\cos 6x \cdot \sin 2x$ as a sum or difference.

Start by plugging in the appropriate product to sum identity:

$12\left(\dfrac{1}{2}\left[\cos\left(6x+2x\right)+\cos\left(6x-2x\right)\right]\right)$. Then simplify using multiplication:

$12\cdot\dfrac{1}{2}\left(\cos 8x+\cos 4x\right)=6\left(\cos 8x+\cos 4x\right)$.

18 Express $2\sin 5x \cdot \cos 2x$ as a sum or difference.

Begin by plugging in the appropriate product to sum identity: $2\left(\dfrac{1}{2}\left[\sin\left(5x+2x\right)+\sin\left(5x-2x\right)\right]\right)$.

Then simplify using multiplication: $2\cdot\dfrac{1}{2}\left(\sin 7x+\sin 3x\right)=\sin 7x+\sin 3x$.

19 Express $6\cos 3x \cdot \sin 6x$ as a sum or difference.

Again, plug in the appropriate product to sum identity: $6\left(\dfrac{1}{2}\left[\sin\left(6x+3x\right)-\sin\left(6x-3x\right)\right]\right)$. Then

simplify using multiplication: $6\cdot\dfrac{1}{2}\left(\sin 9x+\sin 3x\right)=3\left(\sin 9x+\sin 3x\right)$.

20 Express $7\sin 8x \cdot \sin 3x$ as a sum or difference.

You guessed it! Plug in the appropriate product to sum identity:

$7\left(\dfrac{1}{2}\left[\cos\left(8x-3x\right)-\cos\left(8x+3x\right)\right]\right)$. Simplify using multiplication:

$7\cdot\dfrac{1}{2}\left(\cos 5x-\cos 11x\right)=\dfrac{7}{2}\left(\cos 5x-\cos 11x\right)$.

21 Find $\sin 195° - \sin 75°$. **The answer is** $-\dfrac{\sqrt{6}}{2}$.

Begin by using the sum to product identity to rewrite the expression:

$2\cos\left(\dfrac{195+75}{2}\right)\sin\left(\dfrac{195-75}{2}\right)=2\cos 135°\cdot\sin 60°$. Simplify the result using unit circle values:

$2\left(-\dfrac{\sqrt{2}}{2}\right)\left(\dfrac{1}{2}\right)=\cancel{2}\left(-\dfrac{\sqrt{2}}{\cancel{2}}\right)\left(\dfrac{\sqrt{3}}{2}\right)=-\dfrac{\sqrt{6}}{2}$.

22 Find $\cos 375° - \cos 75°$. **The answer is** $\frac{\sqrt{2}}{2}$.

Use the sum to product identity to rewrite the expression:

$-2\sin\left(\frac{375 + 75}{2}\right)\sin\left(\frac{375 - 75}{2}\right) = -2\sin 225° \cdot \sin 150°$. Then, plug in the unit circle values and

simplify: $-2\left(-\frac{\sqrt{2}}{2}\right)\left(\frac{1}{2}\right) = \cancel{2}\left(\frac{\sqrt{2}}{\cancel{2}}\right)\left(\frac{1}{2}\right) = \frac{\sqrt{2}}{2}$.

23 Find $\sin^{7\pi}\!/_{12} + \sin^{\pi}\!/_{12}$. **The answer is** $\frac{\sqrt{6}}{2}$.

Rewrite the expression with the appropriate sum to product identity:

$2\sin\left(\dfrac{\frac{7\pi}{12} + \frac{\pi}{12}}{2}\right)\cos\left(\dfrac{\frac{7\pi}{12} - \frac{\pi}{12}}{2}\right) = 2\sin\left(\dfrac{\frac{8\pi}{12}}{2}\right)\cos\left(\dfrac{\frac{6\pi}{12}}{2}\right) = 2\sin\left(\frac{8\pi}{24}\right)\cos\left(\frac{6\pi}{24}\right) = 2\sin\left(\frac{\pi}{3}\right)\cos\left(\frac{\pi}{4}\right)$.

Use unit circle values to simplify the result: $2\left(\frac{\sqrt{3}}{2}\right)\left(\frac{\sqrt{2}}{2}\right) = \cancel{2}\left(\frac{\sqrt{3}}{\cancel{2}}\right)\left(\frac{\sqrt{2}}{2}\right) = \frac{\sqrt{6}}{2}$.

24 Find $\cos^{23\pi}\!/_{12} + \cos^{5\pi}\!/_{12}$. **The answer is** $\frac{\sqrt{6}}{2}$.

Again, start by using the sum to product identity to rewrite the expression:

$2\cos\left(\dfrac{\frac{23\pi}{12} + \frac{5\pi}{12}}{2}\right)\cos\left(\dfrac{\frac{23\pi}{12} - \frac{5\pi}{12}}{2}\right) = 2\cos\left(\dfrac{\frac{28\pi}{12}}{2}\right)\cos\left(\dfrac{\frac{18\pi}{12}}{2}\right) = 2\cos\left(\frac{28\pi}{24}\right)\cos\left(\frac{18\pi}{24}\right) = 2\cos\left(\frac{7\pi}{6}\right)\cos\left(\frac{3\pi}{4}\right)$.

Next, simplify the result using unit circle values: $2\left(-\frac{\sqrt{3}}{2}\right)\left(-\frac{\sqrt{2}}{2}\right) = \cancel{2}\left(-\frac{\sqrt{3}}{\cancel{2}}\right)\left(-\frac{\sqrt{2}}{2}\right) = \frac{\sqrt{6}}{2}$.

25 Prove $\cos^2 3x - \sin^2 3x = \cos 6x$.

Working on the left side, use the power-reducing formulas to rewrite both terms:

$\left[\frac{1 + \cos 2(3x)}{2}\right] - \left[\frac{1 - \cos 2(3x)}{2}\right] = \cos 6x$. Combine the fractions into one:

$\frac{(1 + \cos 6x) - (1 - \cos 6x)}{2} = \cos 6x$. Combine like terms: $\frac{2\cos 6x}{2} = \cos 6x$. Reduce to complete

the proof: $\cos 6x = \cos 6x$.

26 Prove $(1 + \cos 2x) \cdot \tan^3 x = \tan x(1 - \cos 2x)$.

Start by rewriting the tangent term so that you can use a power-reducing formula:

$(1 + \cos 2x) \cdot \tan x \cdot \tan^2 x = \tan x(1 - \cos 2x)$. Now use the power-reducing formula:

$(1 + \cos 2x) \cdot \left(\frac{1 - \cos 2x}{1 + \cos 2x}\right) \cdot \tan x$. Cancel terms in the numerator and denominator:

$\cancel{(1 + \cos 2x)} \cdot \left(\frac{1 - \cos 2x}{\cancel{1 + \cos 2x}}\right) \cdot \tan x$, and you're there! $\tan x(1 - \cos 2x) = \tan x(1 - \cos 2x)$.

27 Express $\sin^4 x - \cos^4 x$ without exponents. **The answer is $-\cos 2x$.**

For this one, we start by factoring the difference of two squares: $(\sin^2 x + \cos^2 x)(\sin^2 x - \cos^2 x)$. Now use Pythagorean identities (see Chapter 8) to simplify: $(1)(\sin^2 x - \cos^2 x)$. If you replace the cosine term using the same Pythagorean identity, you can combine like terms to have only one squared term remaining: $\sin^2 x - (1 - \sin^2 x) = \sin^2 x - 1 + \sin^2 x = 2\sin^2 x - 1$. From here, you just need to replace the squared term using power-reducing formulas and simplify:

$$2\left(\frac{1 - \cos 2x}{2}\right) - 1 = \cancel{2}\left(\frac{1 - \cos 2x}{\cancel{2}}\right) - 1 = 1 - \cos 2x - 1 = -\cos 2x.$$

28 Express $\cot^2(\frac{x}{2})$ without exponents. **The answer is $\dfrac{1 + \cos x}{1 - \cos x}$.**

Because you don't have a power-reducing formula for cotangent, you need to start by using ratio identities (see Chapter 8) to rewrite the expression: $\dfrac{1}{\tan^2\left(\dfrac{x}{2}\right)}$. Now you can use

tangent's power-reducing formula: $\dfrac{1}{\dfrac{1 - \cos 2\left(\dfrac{x}{2}\right)}{1 + \cos 2\left(\dfrac{x}{2}\right)}}$. Simplify the complex fraction by multiplying

by its reciprocal: $\dfrac{1 + \cos 2\left(\dfrac{x}{2}\right)}{1 - \cos 2\left(\dfrac{x}{2}\right)}$. Finally, make cancellations to simplify: $\dfrac{1 + \cos 2\!\left(\dfrac{x}{\cancel{2}}\right)}{1 - \cos 2\!\left(\dfrac{x}{\cancel{2}}\right)} = \dfrac{1 + \cos x}{1 - \cos x}$.

Chapter 10

Solving Oblique Triangles

· ·

In This Chapter

▶ Using the Law of Sines and the Law of Cosines to solve triangles

▶ Solving word problems using oblique triangles

▶ Finding the area of a triangle

· ·

The trigonometry functions sine, cosine, and tangent are great for finding missing sides and angles inside of right triangles. But what happens when a triangle isn't quite right? These types of triangles are known as *oblique triangles* — any kind of triangle that isn't a right triangle. As you can see in Chapter 6, the process of finding all the sides and angles in a triangle is known as *solving the triangle.* This chapter helps you figure out that process for oblique triangles.

As long as you know one angle and the side directly across from it, you can use the Law of Sines. The Law of Sines can be used in three different cases: ASA, AAS, and SSA. The first two cases have exactly one solution. The third case is known as *the ambiguous case,* as it may have one, two, or no solutions. We take a look at each case to show you how to deal with them. In fact, the ambiguous case gets its own section in this chapter.

If you don't know an angle and the side opposite it, then you start off with the Law of Cosines, which can be used for two cases: SSS and SAS.

 Most books use *standard notation* to label an oblique triangle: Each vertex is labeled with a capital letter, and the side opposite it is the same lowercase letter (across from angle A is side *a,* and so on). We recommend drawing out the triangle and labeling the information you're given. You'll know right away which case you've got on your hands and therefore which formula to use.

Solving a Triangle with the Law of Sines: ASA and AAS

After you draw out your triangle and see that you have two angles and the side in between them (ASA) or two angles and a consecutive side (AAS), proceed to solve the triangle with the Law of Sines:

$$\frac{a}{\sin A} = \frac{b}{\sin B} = \frac{c}{\sin C}$$

In either case, because you know two angles inside the given triangle, you automatically know the third, as the sum of the angles inside any triangle is always 180°. Remember, you'll find exactly one solution in either of these cases of the Law of Sines.

Q. Solve the triangle if A = 54°, B = 28°, and c = 11.2.

A. **C = 98°, a = 9.15, b = 5.31.** First draw and label the triangle, as shown in Figure 10-1. You can see that you have a side sandwiched between two angles (ASA), so you know you can start with the Law of Sines. Because you know two of the angles in the triangle, you can find the third angle first: 54° + 28° + C = 180°; C = 98°. Now substitute all the given information and the angle you just found into the Law of Sines:

$\frac{a}{\sin 54} = \frac{b}{\sin 28} = \frac{11.2}{\sin 98}$. Using the second two equivalent ratios gives you a proportion you can solve: $\frac{b}{\sin 28} = \frac{11.2}{\sin 98}$. Cross multiply to get $b \cdot \sin 98 = 11.2 \cdot \sin 28$. Divide both sides by sin98 to get

$b = \frac{11.2}{\sin 98} \cdot \sin 28$. This goes directly into your calculator to give you b = 5.31.

Note: If you put trig function values into your calculator and round as you go, your final answer will be affected. If, however, you wait until the end like we did in solving for b above, the answer will be more precise.

Set the first and third ratios equal to each other to get the proportion that can be solved for a: $\frac{a}{\sin 54} = \frac{11.2}{\sin 98}$. Cross multiply to get $a \cdot \sin 98 = 11.2 \cdot \sin 54$. Divide the sin98 from both sides to get $a = \frac{11.2}{\sin 98} \cdot \sin 54$ or a = 9.15.

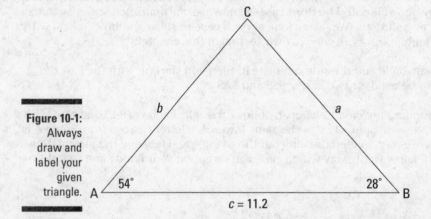

Figure 10-1: Always draw and label your given triangle.

1. Solve the triangle if B = 46°, C = 62°, and a = 21.

Solve It

2. Solve the triangle if A = 19°, C = 100°, and b = 4.4.

Solve It

3. Solve the triangle if A = 49°, B = 21°, and a = 5.

Solve It

4. Solve the triangle if A = 110°, C = 56°, and a = 8.

Solve It

Tackling Triangles in the Ambiguous Case: SSA

A triangle with two sides and a consecutive angle is the *ambiguous case* of the Law of Sines. In this case, you may have one, two, or no solutions. We think it's easiest to assume that two solutions exist. That way, in attempting to find them both, you'll either

Get an error message on your calculator that tells you that there are no solutions

Or you'll find one solution and then

In finding the second one you'll get an answer that doesn't make sense (meaning there really is one solution)

Or, you'll find a second one that works (so there really are two solutions)

When using your calculator to solve these types of problems, anytime you use an inverse trig function (like $\sin^{-1}$, $\cos^{-1}$, or $\tan^{-1}$) to solve for an angle, know that the calculator gives you the reference angle, or the first quadrant answer θ (for more review on this, check out Chapter 6). There's also a second quadrant answer, $180 - \theta$.

Q. Solve the triangle if $a = 25$, $c = 15$, and $C = 40°$.

A. **No solution.** Start off by substituting the given information into the Law of Sines: $\dfrac{25}{\sin A} = \dfrac{b}{\sin B} = \dfrac{15}{\sin 40}$. Notice that the middle ratio has absolutely no information in it at all, so you basically ignore it for now and work with the first and third ratios to get the proportion: $\dfrac{25}{\sin A} = \dfrac{15}{\sin 40}$.

Now, cross-multiply to get $15 \cdot \sin A = 25 \cdot \sin 40$. Solve for the trig function with the variable in it to get $\sin A = \dfrac{25 \sin 40}{15}$. Put the expression into your calculator to get that $\sin A = 1.07$. Even if you forget that sine only has values between -1 and 1 inclusively and you inverse sine both sides of the equation, you get an error message on your calculator that tells you there is no solution.

5. Solve the triangle if $b = 8$, $c = 14$, and $C = 37°$.

Solve It

6. Solve the triangle if $b = 5$, $c = 12$, and $B = 20°$.

Solve It

7. Solve the triangle if $a = 10$, $c = 24$, and $A = 102°$.

Solve It

8. Solve the triangle if $b = 10$, $c = 24$, and $B = 20°$.

Solve It

Conquering a Triangle with the Law of Cosines: SAS and SSS

The Law of Cosines comes in handy when the Law of Sines doesn't work. Specifically, you use the Law of Cosines in two cases:

- ✔ You know two sides and the angle in between them (SAS)
- ✔ You know all three sides (SSS)

The Law of Cosines is:

$$a^2 = b^2 + c^2 - 2bc\cos A$$
$$b^2 = a^2 + c^2 - 2ac\cos B$$
$$c^2 = a^2 + b^2 - 2ab\cos C$$

You may also see your textbook present three forms for each of the angles in the triangle:

$$A = \cos^{-1}\left(\frac{b^2 + c^2 - a^2}{2bc}\right)$$

$$B = \cos^{-1}\left(\frac{a^2 + c^2 - b^2}{2ac}\right)$$

$$C = \cos^{-1}\left(\frac{a^2 + b^2 - c^2}{2ab}\right)$$

You don't have to memorize all these formulas. The ones for the angles are just the first three, each rewritten in terms of the angle. We show the steps for one of them in *Pre-Calculus For Dummies,* so if you'd like to see exactly how you arrive at the angle formulas, check it out there.

Q. Solve the triangle if A = 40°, b = 10, and c = 7.

A. **a = 6.46, C = 44.1°, B = 95.9°.** If you draw out the triangle like the one in Figure 10-2, you notice that this time the angle is between two sides (SAS), and you know that you have to use the Law of Cosines. First, $a^2 = 10^2 + 7^2 - 2(10)(7)\cos40$. Put this right into your calculator to get $a^2 = 41.75$, or a = 6.46. Next, continue using the Law of Cosines to solve for angles (you'll avoid that pesky ambiguous case!) and get $7^2 = 6.46^2 + 10^2 - 2(6.46)(10)\cos C$, or 49 = 41.75 + 100 - 129.2cosC. Isolate for the trig function next and get .718 = cosC (this is why you don't *have* to memorize the formulas for the angles!). Inverse cosine both sides of the equation to get C = 44.1°. Now it's easy to find B because the triangle's angles must total 180°. In this case, B = 95.9°.

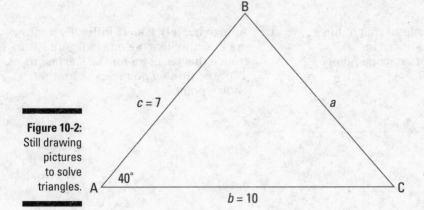

Figure 10-2:
Still drawing pictures to solve triangles.

---

9. Solve the triangle if C = 120°, a = 6, and b = 10.

Solve It

10. Solve the triangle if A = 70°, b = 6, and c = 7.

Solve It

---

11. Solve the triangle if a = 9, b = 5, and c = 7.

Solve It

12. Solve the triangle if a = 7.3, b = 9.9, and c = 16.

Solve It

Using Oblique Triangles to Solve Word Problems

Usually around this time, your textbook or your teacher will present word problems. Argh! Run away, right? Wrong. Draw out a picture and discover that each and every problem at this time is a triangle that isn't right (an *oblique triangle*), where you're looking for one missing piece of information (as opposed to looking for all the angles and sides, like you do in the last three sections). That means less work for you! Each situation requires you to use the Law of Sines or the Law of Cosines exactly once to solve for the missing information.

Q. A plane flies for 300 miles in a straight line, makes a 45° turn, and continues for 700 miles. How far is it from its starting point?

A. **Approximately 936.47 miles.** First, draw out a picture like the one in Figure 10-3. Notice that we use S for the starting point, T for the turning point, and E for the ending point.

Figure 10-3:
A first-class representation of the plane in the problem.

Now that you have the picture, you can figure out whether you need to use the Law of Sines or the Law of Cosines. Because this is SAS, you start off with the modified Law of Cosines, using the different variables from the picture: $t^2 = s^2 + e^2 - 2se\cos T$, or $t^2 = 700^2 + 300^2 - 2(700)(300)\cos 45$. This means that $t^2 = 876,984.85$, and $t = 936.47$.

Q. Two fire towers are exactly 5 miles apart in a forest. They both spot a forest fire, one at an angle of 30° and the other at an angle of 42°. Which tower is closer?

A. **The first fire tower is closer.** Okay, so we lied just a little bit in the introduction to

this section. You aren't always looking for only one missing piece of information. In this problem, you have to find how far both towers are from the fire in order to know which one is closer. But you forgive us, right? First, draw out a figure like the one in Figure 10-4.

Figure 10-4:
Burnin' up
to draw a
picture of
the forest
fire.

This time, we use O for fire tower one, T for fire tower two, and F for the fire itself. You have a classic case of ASA, so you can use the Law of Sines this time. Knowing two of the angles makes it possible to find the third one easily: F = 108°. Now that you have the third angle, you can use the Law of Sines to set up two proportions: $\frac{5}{\sin 108} = \frac{o}{\sin 30}$. Solving for o gets you $o = \frac{5}{\sin 108} \cdot \sin 30$, or o = 2.6 miles.

Now set up another proportion to solve for t: $\frac{5}{\sin 108} = \frac{t}{\sin 42}$, which means that $t = \frac{5}{\sin 108} \cdot \sin 42$, or t = 3.5 miles. That means the first tower is closer.

13. Two trains leave a station at the same time on different tracks that have an angle of 100° between them. If the first train is a passenger train that travels 90 miles per hour and the second train is a cargo train that can only travel 50 miles per hour, how far apart are the two trains after three hours?

Solve It

14. A radio tower is built on top of a hill. The hill makes an angle of 15° with the ground. The tower is 200 feet tall and located 150 feet from the bottom of the hill. If a wire is to connect the top of the tower with the bottom of the hill, how long does the wire need to be?

Solve It

15. A mapmaker stands on one side of a river looking at a flagpole on an island at an angle of 85°. She then walks in a straight line for 100 meters, turns, and looks back at the same flagpole at an angle of 40°. Find the distance from her first location to the flagpole.

Solve It

16. Two scientists stand 350 feet apart, both looking at the same tree somewhere in between them. The first scientist measures an angle of 44° from the ground to the top of the tree. The second scientist measures an angle of 63° from the ground to the top of the tree. How tall is the tree?

Solve It

Figuring Flatness (Area)

In the proof of the Law of Sines, if you go in a slightly different direction you discover a handy formula to find the area of an oblique triangle if you know two sides and the angle between them, as shown in Figure 10-5. The area of the triangle formed is

$$A = \frac{1}{2}ab\sin C.$$

Figure 10-5:
If you know two sides of a triangle and the angle between them, you can calculate the area.

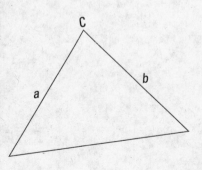

Note: The letters in the formula aren't important. The area of the triangle is *always* one-half the product of the two sides and the sine of the angle between them.

You can also find the area when you know all three sides (*a*, *b*, and *c*) by using what's called Heron's Formula. It says that the area of a triangle is

$$\sqrt{s(s-a)(s-b)(s-c)}, \text{ where } s = \frac{1}{2}(a+b+c)$$

The variable *s* is called the semiperimeter, or half of the triangle's perimeter.

Q. Find the area of the triangle where $b = 4$, $c = 7$, and A = 36°.

A. **The area is about 8.23 square units.** When you have the two sides and the angle between them, you plug the given information into the formula to solve for the area. In this case, A $= \frac{1}{2}(4)(7)\cos 36$, or ≈ 8.23.

17. Find the area of the triangle where $a = 7$, $c = 17$, and B = 68°.

Solve It

18. Find the area of the triangle on the coordinate plane with vertices at (–5, 2), (5, 6), and (4, 0).

Solve It

Answers to Problems on Solving Triangles

1 Solve the triangle if B = 46°, C = 62°, and a = 21. **The answer is A = 72°, b = 15.9, and c = 19.5.**

We draw out this first triangle only, then save some trees and leave the rest to you.

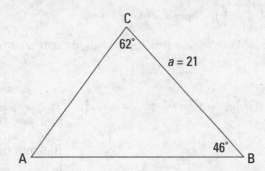

This one is ASA, so you use the Law of Sines to solve it. Because you already know two angles, begin by finding the third: A = 72°. Now set up a proportion from the Law of Sines to solve for the two missing sides. From $\dfrac{21}{\sin 72} = \dfrac{b}{\sin 46}$ you get that $b = \dfrac{21}{\sin 72} \cdot \sin 46$, or b = 15.9. And from $\dfrac{21}{\sin 72} = \dfrac{c}{\sin 62}$ you get that $c = \dfrac{21}{\sin 72} \cdot \sin 62$, or c = 19.5.

2 Solve the triangle if A = 19°, C = 100°, and b = 4.4. **The answer is B = 61°, a = 1.64, and c = 4.95.**

Draw out the figure first. It's ASA again, which means you use the Law of Sines to solve. Find the missing angle first: B = 61°. Now set up the first proportion to solve for a: $\dfrac{a}{\sin 19} = \dfrac{4.4}{\sin 61}$; $a = \dfrac{4.4}{\sin 61} \cdot \sin 19$, or a = 1.64. Set up another proportion to solve for c: $\dfrac{4.4}{\sin 61} = \dfrac{c}{\sin 100}$; $c = \dfrac{4.4}{\sin 61} \cdot \sin 100$, or c = 4.95.

3 Solve the triangle if A = 49°, B = 21°, and a = 5. **The answer is C = 110°, b = 2.37, and c = 6.23.**

This AAS case keeps you on your toes, but you still use the Law of Sines. The missing angle is C = 110°. The first proportion is $\dfrac{5}{\sin 49} = \dfrac{b}{\sin 21}$, which gets you $b = \dfrac{5}{\sin 49} \cdot \sin 21$, or b = 2.37. The second proportion is $\dfrac{5}{\sin 49} = \dfrac{c}{\sin 110}$, which gets you $c = \dfrac{5}{\sin 49} \cdot \sin 110$, or c = 6.23.

4 Solve the triangle if A = 110°, C = 56°, and a = 8. **The answer is B = 14°, b = 2.06, and c = 7.06.**

This one is also AAS, so you use the Law of Sines to solve it. First, the missing angle B is 14°. Now set up the proportion, $\dfrac{8}{\sin 110} = \dfrac{c}{\sin 56}$, to get that $c = \dfrac{8}{\sin 110} \cdot \sin 56$, or c = 7.06. Set up another proportion, $\dfrac{8}{\sin 110} = \dfrac{b}{\sin 14}$, to get that $b = \dfrac{8}{\sin 110} \cdot \sin 14$, or b = 2.06.

5 Solve the triangle if b = 8, c = 14, and C = 37°. **The answer is A = 122.9°, B = 20.1°, and a = 19.53.**

Notice when you draw this one that it's the dreaded SSA, or the ambiguous case. Always assume there are two answers when you're dealing with these types of problems, until you find out otherwise. Set up the proportion $\dfrac{14}{\sin 37} = \dfrac{8}{\sin B}$. This means that $14 \cdot \sin B = 8 \cdot \sin 37$, or $\sin B = \dfrac{8 \sin 37}{14} = 0.344$. Use inverse sine to get that B_1 = 20.1°. This is the first quadrant

answer. The second quadrant has a second answer: $B_2 = 180° - 20.1° = 159.9°$. However, if you look closely you notice that we start off with $C = 37°$. You can't then add a $159.9°$ angle on top of that and still have a triangle, so you throw this second solution away. Only one triangle
satisfies the conditions given. Now that you know C and (the one and only) B, it's easy as pi (Get it? Pi!) to find $A = 122.9°$. Set up another proportion, $\dfrac{14}{\sin 37} = \dfrac{a}{\sin 122.9}$, which means that $a = \dfrac{14}{\sin 37} \cdot \sin 122.9 = 19.53$.

6 Solve the triangle if $b = 5$, $c = 12$, and $B = 20°$. **The answer is $A_1 = 104.8°$, $C_1 = 55.2°$, and $a_1 = 14.13$; or, $A_2 = 35.2°$, $C_2 = 124.8°$, and $a_2 = 8.42$.**

Two solutions! How did we get that? Start at the beginning (a *very* good place to start) and use the Law of Sines to set up the proportion $\dfrac{5}{\sin 20} = \dfrac{12}{\sin C}$. By cross-multiplying, you get the equation $5 \cdot \sin C = 12 \cdot \sin 20$. Solve for sinC by dividing the 5: $\dfrac{12 \sin 20}{5} \approx 0.821$. If sinC $\approx$ 0.821, then use the inverse sine function to discover that $C_1 = 55.2°$. The second quadrant answer is $C_2 = 180 - 55.2 = 124.8°$. If you add $20°$ to *both* of these answers, you discover that it's possible to make a triangle in both cases (because you haven't exceeded $180°$). This sends you on two different paths for two different triangles. We separate them out to make sure you follow the steps.

If $C_1 = 55.2°$, then $A_1 = 104.8°$. Next, set up the proportion $\dfrac{5}{\sin 20} = \dfrac{a_1}{\sin 104.8}$. This means that $a_1 = \dfrac{5}{\sin 20} \cdot \sin 104.8 = 14.13$.

If $C_2 = 124.8°$, then $A_2 = 35.2°$. Set up another proportion, $\dfrac{5}{\sin 20} = \dfrac{a_2}{\sin 35.2}$, to then get that $a_2 = \dfrac{5}{\sin 20} \cdot \sin 35.2 = 8.42$.

7 Solve the triangle if $a = 10$, $c = 24$, and $A = 102°$. **The answer is no solution.**

Here we go again. By the time you're done with this section of pre-calc, you'll be an expert (whether you like it or not) at solving triangles. If you draw this one out, you see another ambiguous SSA case. Set up the proportion $\dfrac{10}{\sin 102} = \dfrac{24}{\sin C}$ using the Law of Sines. Cross-multiply to get $10 \cdot \sin C = 24 \cdot \sin 102$, and then divide the 10 from both sides to get $\sin C = \dfrac{24 \sin 102}{10} \approx 2.35$. That's when the alarms go off. Sine can't have a value bigger than 1, so there's no solution.

8 Solve the triangle if $b = 10$, $c = 24$, and $B = 20°$. **The answer is $A_1 = 104.8°$, $C_1 = 55.2°$, and $b_1 = 28.3$; or, $B_2 = 35.2°$, $C_2 = 124.8°$, and $b_2 = 16.8$.**

Watch as we work it out for you:

$\dfrac{10}{\sin 20} = \dfrac{24}{\sin C}$

$10 \cdot \sin C = 24 \cdot \sin 20$

$\sin C = \dfrac{24 \sin 20}{10} \approx 0.821$

$C_1 = \sin^{-1}(0.821) = 55.2°$

$A_1 = 104.8°$

$\dfrac{10}{\sin 20} = \dfrac{b_1}{\sin 104.8}$

$10 \cdot \sin 101.25 = b_1 \cdot \sin 20$

$\dfrac{10}{\sin 20} \cdot \sin 104.8 = b_1$

$b_1 = 28.3$

$C_2 = 124.8°$ (this solution also works)

$B_2 = 35.2°$

$\dfrac{10}{\sin 20} = \dfrac{b_2}{\sin 35.2}$

$10 \cdot \sin 35.7 = b_2 \cdot \sin 20$

$\dfrac{10}{\sin 20} \cdot \sin 35.2 = b_2$

$b_2 = 16.8$

9 Solve the triangle if C = 120°, $a = 6$, and $b = 10$. **The answer is $c = 14$, A = 21.8°, and B = 38.2°.**

This is definitely a Law of Cosines problem when you draw out the triangle (SAS). Find c first: $c^2 = a^2 + b^2 - 2ab\cos C$. Plug in what you know: $c^2 = 6^2 + 10^2 - 2(6)(10)\cos 120$. Plug this right into your calculator to get that $c^2 = 196$, or $c = 14$. Now find A: $a^2 = b^2 + c^2 - 2bc\cos A$. Plug in what you know: $6^2 = 10^2 + 14^2 - 2(10)(14)\cos A$. Simplify: $36 = 296 - 280\cos A$. Solve for $\cos A$ by first subtracting 296 from both sides: $-260 = -280\cos A$. Now divide the -280 to get $0.929 = \cos A$. Now use inverse cosine to get A = 21.8, and then use the fact that a triangle is 180° to figure out that B = 38.2°.

10 Solve the triangle if A = 70°, $b = 6$, and $c = 7$. **The answer is $a = 7.50$, B = 48.7°, and C = 61.3°.**

By plugging what you know into the Law of Cosines, $a^2 = b^2 + c^2 - 2bc\cos A$, you get $a^2 = 6^2 + 7^2 - 2(6)(7)\cos 70$. This simplifies to $a^2 = 56.27$, or $a = 7.50$.

Now switch the substituting in the second law: $b^2 = a^2 + c^2 - 2ac\cos B$; $6^2 = 7.50^2 + 7^2 - 2(7.50)(7)\cos B$. Simplify: $36 = 105.27 - 105\cos B$. Isolate $\cos B$ by first subtracting 105.27: $-69.27 = -105\cos B$, and then dividing the -105: $0.650 = \cos B$. This means that B = 48.7°. From there you can figure out that C = 61.3°.

11 Solve the triangle if $a = 9$, $b = 5$, and $c = 7$. **The answer is A = 95.7°, B = 33.6°, and C = 50.7°.**

You're solving an SSS triangle using the Law of Cosines, so let's get crackin' to find A:

$a^2 = b^2 + c^2 - 2bc\cos A$

$9^2 = 5^2 + 7^2 - 2(5)(7)\cos A$

$81 = 74 - 70\cos A$

$7 = -70\cos A$

$-0.1 = \cos A$

$95.7° = A$

And again to find B:

$b^2 = a^2 + c^2 - 2ac\cos B$

$5^2 = 9^2 + 7^2 - 2(9)(7)\cos B$

$25 = 130 - 126\cos B$

$-105 = -126\cos B$

0.833 = cosB

33.6° = B

Last, but certainly not least, C = 50.7°.

12 Solve the triangle if a = 7.3, b = 9.9, and c = 16. **The answer is A = 18.3°, B = 25.3°, and C = 136.4°.**

Here we go again to find A:

$a^2 = b^2 + c^2 - 2bc\cos A$

$7.3^2 = 9.9^2 + 16^2 - 2(9.9)(16)\cos A$

$53.29 = 98.01 + 256 - 316.8\cos A$

$-300.72 = -316.8\cos A$

$0.949 = \cos A$

$18.3° = A$

And once more to find B:

$b^2 = a^2 + c^2 - 2ac\cos B$

$9.9^2 = 7.3^2 + 16^2 - 2(7.3)(16)\cos B$

$98.01 = 53.29 + 256 - 233.6\cos B$

$-211.28 = -233.6\cos B$

$0.904 = \cos B$

$25.3° = B$

And finally, C = 136.4°.

13 Two trains leave a station at the same time on different tracks that have an angle of 100° between them. If the first train is a passenger train that travels 90 miles per hour and the second train is a cargo train that can only travel 50 miles per hour, how far apart are the two trains after three hours? **The answer is approximately 330.86 miles apart.**

The two trains depart (F and S) from the same station (T) in a picture like this one:

Notice that you have to use the Law of Cosines to solve for how far apart the two trains are, t.

Next, you need to find how far the trains have traveled to know how far apart they are.

Using D = *rt*, for the first train you get D = 90[3] = 270 miles; for the second train, D = 50[3] = 150 miles. Plug these values into the equation:

$t^2 = f^2 + s^2 - 2fs\cos T$

$t^2 = 150^2 + 270^2 - 2(150)(270)\cos 100$. This goes right into your calculator to give you $t^2 = 109,465.50$, or $t = 330.86$ miles.

14 A radio tower is built on top of a hill. The hill makes an angle of 15° with the ground. The tower is 200 feet tall and located 150 feet from the bottom of the hill. If a wire is to connect the top of the tower with the bottom of the hill, how long does the wire need to be? **The answer is about 279.3 feet long.**

This time our picture is:

To find ∠H in the picture, you add in a horizontal line that's parallel to the ground. Then, using the facts that alternate interior angles are congruent and that the tower has to be completely vertical (or else we have a leaning tower), we know that H = 15° + 90° = 105°.

Now, jump in with the Law of Cosines:

$h^2 = t^2 + b^2 - 2tb\cos H$

$h^2 = 150^2 + 200^2 - 2(150)(200)\cos 105$

$h^2 = 78,029.14$

$h = 279.3$ feet.

15 A mapmaker stands on one side of a river looking at a flagpole on an island at an angle of 85°. She then walks in a straight line for 100 meters, turns, and looks back at the same flagpole at an angle of 40°. Find the distance from her first location to the flagpole. **The answer is 78.5 meters.**

Looking down on the surveyor and the flagpole, here's the picture you use to solve this problem:

Because you have two angles, you can find that P = 55° and use the Law of Sines.

$\dfrac{100}{\sin 55} = \dfrac{s}{\sin 40}$ tells you that $s = \dfrac{100}{\sin 55} \cdot \sin 40 = 78.5$ meters.

16 Two scientists stand 350 feet apart, both looking at the same tree somewhere in between them. The first scientist measures an angle of 44° from the ground to the top of the tree. The second scientist measures an angle of 63° from the ground to the top of the tree. How tall is the tree? **The answer is 226.53 feet tall.**

This problem takes some work. You have to know the distance from either scientist to the top of the tree (FT or TS in the following figure) to know how tall the tree (TB) really is. Here's a drawing of the two scientists and the tree in between them:

Knowing two angles gets you the third one: T = 73°. Law of Sines it is, then!

$\dfrac{350}{\sin 73} = \dfrac{FT}{\sin 44}$ gives you that $FT = \dfrac{350}{\sin 73} \cdot \sin 44 = 326.10$ feet.

Now, because the tree grows straight up, you have a right triangle in which you know one angle and one side. That means you need to go back to SOHCAHTOA (see Chapter 6 for more information) to solve for the missing side, TB. Knowing that you have the hypotenuse and are

looking for the opposite side, you can use the sine function and get $\sin 44 = \dfrac{TB}{254.24}$. This means that TB = 226.53 feet tall. That's one big tree!

17 Find the area of the triangle where $a = 7$, $c = 17$, and B = 68°. **The answer is 55.17.**

Knowing two sides and the angle between them makes this an easy one:

$A = \dfrac{1}{2}(7)(17)\sin 68 = 55.17$.

18 Find the area of the triangle on the coordinate plane with vertices at (–5, 2), (5, 6), and (4, 0). **The answer is 28.**

To figure this out you have to know how long two of the sides are and the angle between them. The problem is, the easiest way to find the angle is to find the length of all three sides and then use the Law of Cosines to find an angle. So you should start by drawing a picture:

Find all three sides first (for a review of how to find the distance between two points, see Chapter 1):

$AC = \sqrt{(5--5)^2+(6-2)^2} = \sqrt{(10)^2+(4)^2} = \sqrt{100+16} = \sqrt{116}$ (this is the same as side a when you use the Law of Cosines)

$AB = \sqrt{(4--5)^2+(8-6)^2} = \sqrt{(9)^2+(2)^2} = \sqrt{81+4} = \sqrt{85}$ (this is side c)

$BC = \sqrt{(4-5)^2+(0-6)^2} = \sqrt{(-1)^2+(-6)^2} = \sqrt{1+36} = \sqrt{37}$ (this is side b)

Now that you've found the length of all three sides, use Heron's Formula to find the area:

$s = \frac{1}{2}\left(\sqrt{116} + \sqrt{85} + \sqrt{37}\right) \approx 13.04.$

$A = \sqrt{13.04(13.04-10.77)(13.04-9.22)(13.04-6.08)} = \sqrt{13.04(2.27)(3.82)(6.96)} = \sqrt{787} = \approx 28.05.$

Part IV
And the Rest . . .

In this part . . .

This is the section for everything else. The topics aren't
necessarily related other than the fact that they're all
covered in a typical Pre-Calc course. Watch out for those
unsignaled left turns!

This part begins with how to perform operations with and
graph complex numbers. We also introduce the idea of
polar coordinates, a brand new way of graphing equations!
We know you've been asking yourself when that was
gonna happen . . . well, here you go. Conic sections are
also a great thing to graph, so we cover them in detail.

We then move on to the systems of equations and cover
solving linear and nonlinear equations, as well as working
with matrices. Next, we're on to sequences and series. We
discuss how to find any term in a sequence, how to calculate the sum of a sequence, and how to write the formula
that determines a given sequence. Lastly, we cover the
topics that usually constitute the end of pre-calc (and the
beginning of calc): limits and continuity.

Chapter 11

Complex Numbers and Polar Coordinates

*O*nce upon a time, mathematicians delved into their imaginations and invented a whole new set of numbers. They were so deep into their imaginations that they decided to call these numbers imaginary numbers, because there was just no way these new numbers would ever pop up in the real world.

Well, they were wrong. These imaginary numbers did eventually appear. Fields like engineering, electricity, and quantum physics all use imaginary numbers in their everyday applications. An *imaginary number* is basically the square root of a negative number. The *imaginary unit*, denoted i, is the solution to the equation $i^2 = -1$.

A *complex number* is the sum of a real number and an imaginary one. $a + bi$ is an example of a complex number, where a is the real part and bi is the imaginary part. Pure real numbers like 17 are considered complex numbers as well, but the imaginary part is 0 (it can be written as $17 + 0i$). Pure imaginary numbers like $-2i$ are similar, but the real part is 0 (it can be written as $0 - 2i$).

This chapter is like the bonus features on your favorite DVD, as it also includes graphing points and equations in a whole new way, called "polar coordinates." Hang in there for a bit while we explore complex numbers in depth before moving onto polar coordinates.

Performing Operations with and Graphing Complex Numbers

Complex numbers in the form $a + bi$ can be graphed on a *complex coordinate plane.* Each real number is placed on the horizontal axis (so now it's called the *real axis*), and each imaginary number is placed on the vertical one (called the *imaginary axis*). Each complex number $a + bi$ corresponds to a point (a, b) on the complex plane. Figure 11-1 shows several examples of points on the complex plane.

Figure 11-1:
Graphing
complex
numbers.

Adding and subtracting complex numbers is just another example of collecting like terms: You can only add or subtract real numbers and you can only add or subtract imaginary numbers.

When multiplying complex numbers, you FOIL the two binomials. All you have to do is remember that the imaginary unit is defined such that $i^2 = -1$, so any time you see i^2 in an expression, replace it with -1. When dealing with other powers of i, notice the pattern here:

$$i = \sqrt{-1} \qquad i^5 = i$$
$$i^2 = -1 \qquad i^6 = -1$$
$$i^3 = -i \qquad i^7 = -i$$
$$i^4 = 1$$

It continues in this manner forever, repeating in a cycle every fourth power. To find a larger power of i, rather than counting forever, realize that the pattern repeats. For example, to find i^{243}, divide 4 into 243 and you get 60 with a remainder of 3. It will repeat the pattern 60 times and then have 3 left over, so i^{243} is the same as i^3, which is $-i$.

The *conjugate* of a complex number $a + bi$ is $a - bi$, and vice versa. When you multiply two complex numbers that are conjugates of each other, you always end up with a pure real number:

$(a + bi)(a - bi)$.

FOIL the binomials: $a^2 - abi + abi - b^2i^2$.

Cancel the two middle terms: $a^2 - b^2i^2$.

Replace i^2 with -1: $a^2 - b^2(-1)$.

Simplify: $a^2 + b^2$.

When dividing complex numbers, you end up with a root in the denominator (because if $i^2 = -1$, then $i = \sqrt{-1}$). This means that you have to rationalize the denominator. To do so, you must multiply the complex number in the denominator by its conjugate, and then multiply this same expression in the numerator. (For more information on rationalizing the denominator of any fraction, see Chapter 2.)

We show you how to do this in variables, but you may like it better if you go to the forthcoming example to see how to rationalize a fraction with complex numbers involved.

$$\frac{a+bi}{c+di}.$$

Multiply both the numerator and the denominator by the conjugate of the denominator: $\frac{a+bi}{c+di} \cdot \frac{c-di}{c-di}.$

FOIL the numerators and the denominators: $\frac{ac-adi+bci-bdi^2}{c^2-cdi+cdi-d^2i^2}.$

Cancel like terms in the denominator and replace the i^2 with -1 in the numerator and denominator: $\frac{ac-adi+bci+bd}{c^2+d^2}.$

Q. Perform the indicated operation: $(3-4i) + (-2+5i)^2.$

A. **$-18-24i$.** Follow your order of operations (PEMDAS) and square the second binomial by FOILing it times itself: $(3-4i) + (-2+5i)(-2+5i)$; $(3-4i) + (4-10i-10i+25i^2)$. Combine the like terms: $(3-4i) + (4-20i+25i^2)$. Substitute -1 for i^2: $(3-4i) + (4-20i-25)$. Combine like terms: $(3-4i) + (-21-20i)$. Combine like terms one more time — reals with reals and imaginary with imaginary: $-18-24i$.

Q. Perform the indicated operation: $\frac{6}{2-9i}.$

A. $\frac{12+54i}{85}$. The conjugate of the denominator is $2+9i$. Multiply this on the top and bottom of the fraction: $\frac{6}{2-9i} \cdot \frac{2+9i}{2+9i}$. Distribute the numerator and FOIL the denominator: $\frac{12+54i}{4+18i-18i-81i^2}$. Cancel like terms in the denominator and replace i^2 with -1: $\frac{12+54i}{4+81}$. This simplifies to the answer $\frac{12+54i}{85}$.

Q. Graph $4-6i$ on the complex coordinate plane.

A. **See the following graph.** Go to the right 4 units on the real axis and 6 units down on the imaginary axis and place a point. Easy!

Q. Find x and y: $3x + 4yi = 6 - 2i$.

A. **$x = 2$, $y = -\frac{1}{2}$.** The real parts of both sides of this equation must equal each other: $3x = 6$. Divide both sides by 3 to get the solution $x = 2$. Furthermore, the imaginary parts must also equal each other: $4yi = -2i$, so that when you divide both sides by $4i$ and reduce, you get $y = -\frac{1}{2}$.

1. Plot the point $4 - 3i$ on the complex plane.

Solve It

2. Find $|3 - 4i|$.

Solve It

3. Find i^{22}.

Solve It

4. Solve the equation $5x^2 - 2x + 3 = 0$.

Solve It

Round a Pole: Graphing Polar Coordinates

Up until now in your math career, you've been graphing everything based on the *rectangular coordinate system*. It's called that because it's based on two number lines perpendicular to each other. Pre-calc takes that concept further when it introduces *polar coordinates*.

In polar coordinates, every point is located around a central point, called the *pole,* and is named (r, θ). r is the radius, and θ is the angle formed between the polar axis (think of it as what *used* to be the positive *x*-axis) and the segment connecting the point to the pole (what *used* to be the origin).

Most books use radians when measuring angles in polar coordinates, so we do the same thing. If you need a recap on radians, see Chapter 6. Figure 11-2 shows the polar coordinate plane.

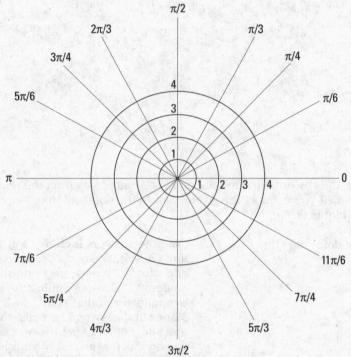

Figure 11-2:
Graphing
round and
round on
the polar
coordinate
plane.

Notice that a point on the polar coordinate plane has more than one name. Because you're moving in a circle, you can always add or subtract 2π to any angle and end up at the same point. This concept is so important in graphing equations in polar forms that we dedicate this entire section to making sure that you understand.

When both the radius and the angle are positive, most of our students have no difficulties finding the point on the polar plane. If the radius is positive and the angle is negative, the point moves in a clockwise direction, just like radians do. If the radius is negative and the angle is positive, find the point where both are positive first and then reflect that point across the pole. If both the radius and the angle are negative, find the point where the radius is positive and the angle is negative and then reflect that across the pole.

Q. What's the polar coordinate of point P in the following figure?

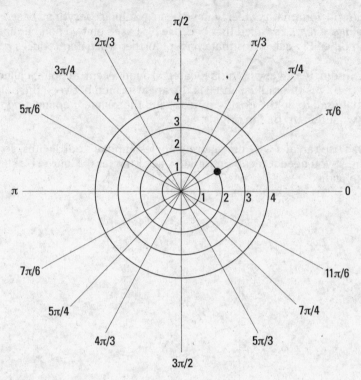

A. **(2, ⅓₄).** First determine the radius by noticing that the point is 2 units away from the pole. Also notice that an angle forms when you connect the given point to the pole and that the polar axis is a ⅓₄ angle. This means the point is (2, ⅓₄).

Q. Name two other points that determine the same point in the last question.

A. **Possible answers include (2, ⁹ᵖ/₄), (2, ¹⁷ᵖ/₄), and (2, −⁷ᵖ/₄).** You add 2π to the angle twice and subtract it once to get these three angles. It's all about finding the common denominator at that point: ⅓₄ + 2π = ⅓₄ + ⁸ᵖ/₄ = ⁹ᵖ/₄, our first answer. Then take that answer and add 2π to it to get the next one. You can do this for the rest of your life and still not list all the possibilities.

5. Graph (4, −⁵⁄₃).

Solve It

6. Graph (−5, ½).

Solve It

7. Name two other polar coordinates for the point in question 5: one with a negative angle and one with a positive angle.

Solve It

8. Name two other polar coordinates for the point in question 6: one with a negative angle and one with a positive radius.

Solve It

Changing to and from Polar

You can use both polar and rectangular coordinates to name the same point on the coordinate plane. Sometimes it's easier to write an equation in one form than the other, but usually, pre-calc books begin by having you switch between the two just to get used to them. Figure 11-3 shows how to determine the relationship between these two not-so-different methods.

Figure 11-3:
A right triangle reveals the relationship between rectangular and polar coordinates.

Some simple right triangle trig and a little Pythagorean Theorem determine the relationship:

- $\sin\theta = \frac{y}{r}$ or $y = r\sin\theta$
- $\cos\theta = \frac{x}{r}$ or $x = r\cos\theta$
- $x^2 + y^2 = r^2$
- $\tan\theta = \frac{y}{x}$ or $\theta = \tan^{-1}(\frac{y}{x})$

Q. Rewrite the equation in rectangular form: $r = 2\sin\theta$.

A. $x^2 + y^2 = 2y$. First, convert all trig functions to x, y, and r. If $r = 2\sin\theta$, then $r = 2 \cdot \frac{y}{r}$. Then multiply the r to the other side to get $r^2 = 2y$. Then you can replace r^2 from the Pythagorean Theorem and get $x^2 + y^2 = 2y$.

Q. Rewrite the equation $r = 2\csc\theta$ in rectangular form.

A. $y = 2$. First, realize that cosecant is the reciprocal of sine (see Chapter 6 for a refresher). If $\sin\theta = \frac{y}{r}$, then $\csc\theta = \frac{r}{y}$. Substitute this into the equation to get $r = 2\frac{r}{y}$.

Multiply the y to the other side and get $ry = 2r$. This means that $y = 2$.

9. Rewrite $r = 3\sin\theta + 4\cos\theta$ in rectangular form.

Solve It

10. Rewrite $y = 2x - 1$ in polar form.

Solve It

11. Rewrite $3x - 5y = 10$ in polar form.

Solve It

12. Rewrite $x^2 + y^2 = 16$ in polar form.

Solve It

13. Convert the polar coordinate (4, ⅚) to rectangular coordinates.

Solve It

14. Convert the rectangular coordinate (–1, 1) to polar coordinates.

Solve It

Graphing Polar Equations

When given an equation in polar coordinates and asked to graph it, most students go with the plug-and-chug method: Pick values for θ from the unit circle that you know so well and find the value of r. Polar equations have various types of graphs, and we take a closer look at each one. Be sure to also see Chapter 12 (conic sections) for information about how to graph conic sections in polar coordinates.

Archimedean spiral

$r = a\theta$ gives a graph that forms a spiral. a is a constant that's multiplying the angle, and then the radius is the same. If a is positive, the spiral moves in a counterclockwise direction, just like positive angles do. If a is negative, the spiral moves in a clockwise direction.

Cardioid

You may recognize the word cardioid if you've ever worked out and done your cardio. The word relates to the heart, and when you graph a cardioid, it does look like a heart, of sorts. Cardioids are written in the form $r = a(1 \pm \sin\theta)$ or $r = a(1 \pm \cos\theta)$. The cosine equations are hearts that point to the left or right, and the sine equations open up or down.

Rose

A rose by any other name is . . . a polar equation. If $r = a\sin n\theta$ or $r = a\cos n\theta$, the graphs look like flowers with petals. The number of petals is determined by n. If n is odd, then there are n (the same number of) petals. If n is even, there are $2n$ petals.

Circle

When $r = a\sin\theta$ or $r = a\cos\theta$, you end up with a circle with a diameter of a. Circles with cosine in them are on the x-axis, and circles with sine in them are on the y-axis.

Lemniscate

A lemniscate makes a figure eight; that's the best way to remember it. $r^2 = \pm a^2\sin 2\theta$ forms a figure eight in between the axes, and $r^2 = \pm a^2\cos 2\theta$ forms a figure eight that lies on one of the axes.

Limaçon

A cardioid is really a special type of limaçon, which is why they look similar to each other when graphing them. The limaçon form is familiar:

$$r = a \pm b \sin\theta$$
$$r = a \pm b \cos\theta$$

Q. Sketch the graph of $r = 1 + \sin\theta$.

A. **See the following graph.** This is a cardioid. If you're lucky enough to have a teacher who lets you use a graphing calculator to graph polar equations, make sure your calculator is set to radians, input the equation into the grapher, and presto, you have yourself a graph. If not, just plug and chug the equation to get the graph. For instance, if $\theta = 0$, then $r = 1 + \sin 0 = 1 + 0 = 1$. If $r = \frac{\pi}{2}$, then $r = 1 + \sin\frac{\pi}{2} = 1 + 1 = 2$. Keep going

in this manner until you end up with the graph.

$r = 1 + \sin\theta$

Q. Sketch the graph of $r = \cos 3\theta$.

A. **See the following graph.** This is a rose with three petals because the coefficient on the inside is odd. Plug and chug this one as well, as shown in the following chart (we show the first quadrant values only):

θ	cos3θ	r
0	cos3(0)	1
$\frac{\pi}{6}$	cos$\frac{\pi}{2}$	0
$\frac{\pi}{4}$	cos3$\frac{\pi}{4}$	$-\sqrt{\frac{2}{2}}$
$\frac{\pi}{3}$	cosπ	−1
$\frac{\pi}{2}$	cos$\frac{3\pi}{2}$	0

r = cos 3θ

15. Sketch the graph of $r2 = 9\cos2q$.

Solve It

16. Sketch the graph of $r = 2\theta$.

Solve It

17. Sketch the graph of $r = 1 - 3\sin\theta$.

Solve It

18. Sketch the graph of $r = 3 + 2\sin\theta$.

Solve It

Answers to Problems on Complex Numbers and Polar Coordinates

1 Plot the point $4 - 3i$ on the complex plane. **See the following graph for the answer.**

The real unit is 4 to the right and the imaginary unit is 3 down. This lands you at the point in the figure.

2 Find $|\,3 - 4i\,|$. **The answer is 5.**

Remember that absolute value bars represent distance. In the case of complex numbers, they represent the distance from the point to the origin. This distance is always the same as the length of the hypotenuse of the right triangle drawn when connecting the point to the x- and y-axes. $d = \sqrt{3^2 + (-4)^2} = \sqrt{9 + 16} = \sqrt{25} = 5$.

3 Find i^{22}. **The answer is –1.**

The pattern for the powers of i repeats every four. Divide 22 by 4 and get 5 with a remainder of 2. This means that $i^{22} = i^2 = -1$.

4 Solve the equation $5x^2 - 2x + 3 = 0$. **The answer is $\dfrac{1 \pm \sqrt{14}i}{5}$.**

This quadratic equation is unfactorable, so you have to resort to the quadratic formula to solve it. $x = \dfrac{2 \pm \sqrt{(-2)^2 - 4(5)(3)}}{2(5)} = \dfrac{2 \pm \sqrt{4 - 60}}{10} = \dfrac{2 \pm \sqrt{-56}}{10} = \dfrac{2 \pm 2\sqrt{14}i}{10}$, which simplifies to $\dfrac{1 \pm \sqrt{14}i}{5}$. If you've forgotten how to deal with the quadratic formula and/or how to simplify roots, see Chapter 4.

5 Graph $(4, -\tfrac{5\pi}{3})$. **See the following graph for the answer.**

The radius is 4 and the angle is negative, which moves in a clockwise direction and ends up in the first quadrant.

6 Graph (–5, ½). **See the following graph for the answer.**

The radius for this one is negative, which reflects the point (5, ½) over the pole.

$(-5, \frac{\pi}{2})$

7 Name two other polar coordinates for the point in question 5: one with a negative angle and one with a positive angle. **Possible answers include (4, –¹¹π/₃), (4, ⁷π/₃).**

The radius isn't changing, so don't do anything to it. Subtract 2π from the angle to get the first answer. Add 2π to the angle to get the second answer.

8 Name two other polar coordinates for the point in question 6: one with a negative angle and one with a positive radius. **Possible answers include (–5, –³π/₂), (5, ³π/₂).**

To deal with the first situation, don't change the radius; just subtract 2π to get a negative angle: –³π/₂. To change the radius to a positive 5, you have to change the angle. Because the point was down 5 originally, this angle is the same as ³π/₂.

9 Rewrite $r = 3\sin\theta + 4\cos\theta$ in rectangular form. **The answer is $x^2 + y^2 = 3y + 4x$.**

First, change $\sin\theta$ to $\frac{y}{r}$ and $\cos\theta$ to $\frac{x}{r}$: $r = 3\frac{y}{r} + 4\frac{x}{r}$. Multiply everything by r to get rid of the fractions and get $r^2 = 3y + 4x$. Use the Pythagorean substitution and get $x^2 + y^2 = 3y + 4x$.

10 Rewrite $y = 2x - 1$ in polar form. **The answer is $r = \dfrac{-1}{\sin\theta - 2\cos\theta}$.**

To move the other direction, remember that $x = r\cos\theta$ and $y = r\sin\theta$. Make these substitutions first and get: $r\sin\theta = 2r\cos\theta - 1$. Get all terms with r to one side first: $r\sin\theta - 2r\cos\theta = -1$. Factor out the common factor of r: $r(\sin\theta - 2\cos\theta)$. Now solve for r by dividing the rest to the other side: $r = \dfrac{-1}{\sin\theta - 2\cos\theta}$.

11 Rewrite $3x - 5y = 10$ in polar form. **The answer is $\dfrac{10}{3\cos\theta - 5\sin\theta}$.**

Use the same substitutions as the last question to get $3r\cos\theta - 5r\sin\theta = 10$. Factor: $r(3\cos\theta - 5\sin\theta) = 10$. Divide: $\dfrac{10}{3\cos\theta - 5\sin\theta}$.

12 Rewrite $x^2 + y^2 = 16$ in polar form. **The answer is $r = 4$.**

Use the Pythagorean substitution first: $x^2 + y^2 = r^2$, so $r^2 = 16$. Take the square root of both sides and get $r = 4$.

13 Convert the polar coordinate (4, π/₆) to rectangular coordinates. **The answer is $(2\sqrt{3}, 2)$.**

If $x = r\cos\theta$, then $x = 4\cos\frac{\pi}{6} = 4\left(\dfrac{\sqrt{3}}{2}\right) = 2\sqrt{3}$.

If $y = r\sin\theta$, then $y = 4\sin\frac{\pi}{6} = 4(½) = 2$.

14 Convert the rectangular coordinate (–1, 1) to polar coordinates. **The answer is ($\sqrt{2}$, –¾).**

Use the Pythagorean substitution first to find r: $r^2 = x^2 + y^2$ means that $r^2 = (-1)^2 + (1)^2 =$ $1 + 1 = 2$, or $r = \sqrt{2}$. Then use tangent to find the angle: $\tan\theta = -1$, or $\theta = -\frac{3}{4}$.

15 Sketch the graph of $r^2 = 9\cos2\theta$. **See the following graph for the answer.**

This graph is a lemniscate. If $\theta = 0$, $r = 3$. If $\theta = \frac{1}{6}$, then $r \approx 2.12$, and so on. Plug and chug it and wind up with the graph.

16 Sketch the graph of $r = 2\theta$. **See the following graph for the answer.**

This is a spiral. If $\theta = 0$, $r = 0$; if $\theta = \frac{1}{6}$, $r \approx 1.14$, and so on.

17 Sketch the graph of $r = 1 - 3\sin\theta$. **See the following graph for the answer.**

This is a limaçon. Plug and chug it and get the graph.

18 Sketch the graph of r = 3 + 2sinθ. **See the following graph for the answer.**

This is a cardioid. If you plug and chug it, you end up with the graph.

r = 3 + 2sinθ

Chapter 12

Conquering Conic Sections

- -

In This Chapter

▶ Closing in on circles

▶ Graphing parabolas, ellipses, and hyperbolas

▶ Recognizing the different conic sections

▶ Working with parametric form and polar coordinates

- -

Who doesn't love a good cone? We're big fans of the ice cream cone ourselves, but maybe you're partial to the traffic cone because it keeps you safe on the road. Whatever type of cone is your favorite, for mathematicians the cone is the creative fuel for the fire of a whole bunch of ideas.

You see, about 2,200 years ago, some smart mathematician named Apollonius of Perga decided one day to stack two cones point to point. He sliced them in different directions and came up with four different *conic sections*: the *circle*, the *ellipse*, the *parabola*, and the *hyperbola*. Each conic section has its own equation and its own parts, which you need to determine in order to graph it. Typically, a textbook question will ask you to graph a conic section, identify certain parts, or write its equation. To do *any* of these tasks, you must be able to recognize what kind of conic section it is (more on that later) and write it in its own equation form.

When a conic section is already written in its form, you know most of the information about the parts of that particular conic. When a conic section isn't written in its form, well, you just have to write it that way. How do you do that, you ask (and we're so glad that you did)? It's the process called *completing the square*. That should sound familiar to you from previous math classes. If not, review Chapter 4.

A Quick Conic Review

To save you some time, we quickly review here how to complete the square for conic sections. This is the *only* way to change a conic section's equation that's not written in its proper form to one that is. The steps of the process are as follows:

1. **Add/subtract any constant to the opposite side of the given equation, away from all the variables.**

2. **Factor the leading coefficient out of all terms in front of the set of parentheses.**

3. **Divide the remaining linear coefficient by two, but only in your head.**

4. **Square the answer from Step 3 and add that inside the parentheses.** Don't forget that if you have a coefficient from Step 2, you must multiply the coefficient and this number from Step 4 and add *that* to both sides.

5. **Factor the quadratic polynomial as a perfect square trinomial.**

This chapter is dedicated to each mathematical conic superstar, one at a time — how to write each one in its form and how to graph it.

Going Round and Round with Circles

A circle is simple, but not so plain. You can do so much with a circle. The very tires you drive on are, after all, circular in shape. A circle has one point in the very middle called the *center*. All the points on the circle are the same distance, called the *radius*, from the center.

When a circle is drawn centered at the origin of the coordinate plane, the equation that describes it is simple as well:

$$x^2 + y^2 = r^2$$

r is the variable that represents the circle's radius.

When the circle is moved around the coordinate plane with horizontal and/or vertical shifts, its equation looks like this:

$$(x - h)^2 + (y - v)^2 = r^2$$

h is the horizontal shift of the equation, v is the vertical shift, and r is still the radius. If a circle isn't written in this form, you'll still recognize it as a circle because the equation will have *both* an x^2 and a y^2 and the coefficients on both will be equal.

Don't be intimidated if you see this written another way in your pre-calc textbook. Different books may write this equation differently. Just know that it's always written as $(x - \text{horizontal})^2$ and $(y - \text{vertical})^2$.

Q. What's the center and the radius of the circle $(x + 3)^2 + y^2 = 16$?

A. **Center (–3, 0); radius r = 4.** Because we've been talking about circles, we thought we'd start you off with training wheels. This equation is already in the proper form of a circle, so finding its information is easy. The horizontal value with x is 3, so $h = -3$; the vertical shift is missing, so $v = 0$. This means that the center is located at the point (–3, 0). Meanwhile, the other side of the equation gives you the radius squared. Setting $r^2 = 16$ gives you the solution $r = 4$.

Q. Graph $x^2 + y^2 - 6y + 2 = 0$.

A. **See Figure 12-1.** Now the training wheels come off! First subtract the constant to move it to the other side: $x^2 + y^2 - 6y = -2$. Because the leading coefficients are 1, you don't have to factor anything out and can move onto the next step. Then, because the x variable doesn't have a linear term, you don't have to complete the square there. Notice, however, that the y variable does have one, so you follow the process: $\left(\dfrac{-6}{2}\right)^2 = 9$. This value gets added to both sides: $x^2 + y^2 - 6y + 9 = 7$. Factor the y's as a perfect square trinomial to get $x^2 + (y - 3)^2 = 7$. *Voilá!* You got yerself a circle. Its center is (0, 3), and its radius is $\sqrt{7}$. See Figure 12-1 for the graph of this beautiful circle.

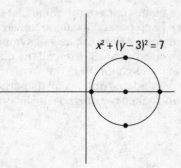

Figure 12-1:
Graphing a circle: Mark the center first, and then mark the radius in all directions.

$x^2 + (y-3)^2 = 7$

1. Find the center and radius of the circle $2x^2 + 2y^2 - 4x = 15$. Then graph the circle.

Solve It

2. Write the equation of the circle with the center $(-1, 4)$ if the circle passes through the point $(3, 1)$.

Solve It

Graphing Parabolas: The Ups and Downs

When you graph a *quadratic polynomial* (see Chapter 4 for more information on this type of polynomial), you always get a parabola. Typically, up until this point in pre-calc, the graphs of parabolas have been vertical: they open up or down. In conic sections, however, they can also open horizontally: to the left or to the right. We take a look at each situation in the following sections.

Officially, a *parabola* is the set of all points on a plane that are the same distance from a given point (*focus*) and a given line (*directrix*). Each parabola can be folded exactly in half over a line called the *axis of symmetry*. The point where the axis of symmetry intersects the graph is called the *vertex*. This gives you what we like to call the martini of conic sections: The parabola is the glass, the axis of symmetry is the stem, the directrix is the base, and the focus is the olive. Every good martini has all its parts, and every good parabola does, too! But don't be too shaken (or stirred) up! Figure 12-2 shows all the parts of the parabola.

Figure 12-2:
The parabola-tini . . . no twist.

Standing tall: Vertical parabolas

The equation of a vertical parabola is:

$$y = a(x - h)^2 + v$$

where h is the horizontal shift, v is the vertical shift, and a is the vertical transformation. You'll recognize a vertical parabola when x is squared but y isn't. We discuss these types of transformations ad nauseum in Chapter 3, and it shouldn't surprise you that nothing changes just because we're calling them conic sections. Each vertical parabola still has the following parts:

- ✔ **Vertex:** (h, v)
- ✔ **Axis of symmetry:** $x = h$
- ✔ **Focus:** $(h, v + \frac{1}{4a})$
- ✔ **Directrix:** $y = v - \frac{1}{4a}$

Don't memorize these as formulas to find the parts. Instead, use what you know about a parabola to get the job done. The vertex is always the first point that you graph on a parabola. From there, know that the focus and the directrix are $\frac{1}{4a}$ away from it. One will be above and one will be below, depending on the value of a. If $a < 0$, the parabola opens down and the focus moves down while the directrix moves up. If $a > 0$, the parabola opens up, the focus moves up, and the directrix moves down.

Q. Graph the equation of the parabola $y = -2(x - 1)^2 + 5$.

A. **See Figure 12-3.** Because this equation is already in the proper parabola form, you should be able to go right to graphing. The vertex is at $(1, 5)$. The vertical transformation is 2, and the graph is turned upside down. If you don't know where this information comes from, we strongly recommend you go back and read Chapter 3 now, which is chock-full of information on transforming any function. We'll be waiting right here for you when you get back. In the meantime, the final graph is shown in Figure 12-3.

Vertex (1, 5)

$y = -2(x - 1)^2 + 5$

Figure 12-3: Graphing a vertical parabola.

Q. State the vertex, axis of symmetry, focus, and directrix of $y = 3x^2 - 4x + 1$.

A. **Vertex: $(\frac{2}{3}, -\frac{1}{3})$; axis of symmetry: $x = \frac{2}{3}$; focus: $(\frac{2}{3}, -\frac{1}{4})$; directrix: $y = -\frac{1}{12}$.** That's an awful lot of fractions, ain't it? But the process doesn't change. Start by subtracting 1 from both sides: $y - 1 = 3x^2 - 4x$. Then factor out the three: $y - 1 = 3(x^2 - \frac{4}{3}x)$.

Now complete the square and be sure to keep the equation balanced: $y - 1 + \frac{4}{3} = 3(x^2 - \frac{4}{3}x + \frac{4}{9})$. Simplify and factor: $y + \frac{1}{3} = 3(x - \frac{2}{3})^2$. Then solve for y to put the parabola in its proper form: $y = 3(x - \frac{2}{3})^2 - \frac{1}{3}$. Next, use the formulas to figure out all the parts — the vertex is $(\frac{2}{3}, -\frac{1}{3})$, the axis of symmetry is $x = \frac{2}{3}$, the focus is $(\frac{2}{3}, -\frac{1}{4})$, and the directrix is $y = -\frac{1}{12}$. Sheesh!

3. What's the vertex of the parabola $y = -x^2 + 4x - 6$? Sketch the graph of this parabola.

Solve It

4. Find the focus and the directrix of the parabola $y = 4x^2$.

Solve It

Lying down on the job: Horizontal parabolas

The equation of a horizontal parabola is very similar to the vertical one we discuss in the last section. Here it is:

$$x = a(y - v)^2 + h$$

This is a horizontal parabola because y is squared but x is not. Notice, also, that h and v are still there for the horizontal and vertical shifts, respectively, but that they've switched places. Because this parabola is horizontal, a also switches to become the horizontal transformation. This is the first time we've talked about a horizontal transformation, so we'll take a few moments to explain the idea. A *horizontal transformation* does the same thing as a vertical transformation, but it affects what the function does from left to right. A horizontal transformation where a is a fraction in between 0 and 1 is a *horizontal shrink,* and a horizontal transformation where $a > 1$ is called a *horizontal stretch.* All of these are positive, so the parabola opens to the right. When a is negative the parabola does the same thing, but the graph is reflected in the opposite direction (so the parabola opens to the left).

Here are the parts of a horizontal parabola:

- ✔ **Vertex:** (h, v)
- ✔ **Axis of symmetry:** $y = v$
- ✔ **Focus:** $(h + \frac{1}{4a}, v)$
- ✔ **Directrix:** $x = h - \frac{1}{4a}$

Q. Graph the parabola if its equation is $x = y^2 - 6y$.

A. **See Figure 12-4.** Complete the square to get this horizontal parabola in its form: $x = (y - 3)^2 - 9$. This means that the parabola's vertex is located at the point $(-9, 3)$. There's no horizontal transformation ($a = 1$), so from the vertex the graph moves up 1, over 1; up 1, over 4; up 1, over 9. This gives you the graph in Figure 12-4.

Figure 12-4:
Graphing a
horizontal
parabola.

Q. Write the equation of the parabola whose vertex is $(-2, 1)$ if the focus is at the point $(-4, 1)$.

A. $x = \dfrac{-1}{8}(y - 1)^2 - 2$. This one takes some brainpower to work out. Because the focus is to the left of the vertex, you know that the parabola is a horizontal one (forgetting, of course, that you're in the section of the book on horizontal parabolas). Start with the equation of any horizontal parabola: $x = a(y - v)^2 + h$. Then plug in the vertex values $(-2, 1)$: $x = a(y - 1)^2 - 2$. Now, all you have to do is figure out what that pesky value of a is. Have no fear, pre-calc is here! You know that the equation to find the focus is $(h + \dfrac{1}{4a}, v)$. This tells you that $h + \dfrac{1}{4a} = -4$. You also know that h is -2, so substitute and get $-2 + \dfrac{1}{4a} = -4$. Solve to get $a = -\frac{1}{8}$. Finally, write the equation of the parabola: $x = \dfrac{-1}{8}(y - 1)^2 - 2$.

5. Sketch the graph of $x = 2(y - 4)^2$.

Solve It

6. Determine whether this parabola opens left or right: $x = -y^2 - 7y + 3$.

Solve It

Graphing Ellipses: The Fat and the Skinny

An *ellipse* is defined as the set of all points on a plane, such that the sum of the distances from any point on the curve to two fixed points, the foci, is a constant. Think of an ellipse as a circle that's gone flat, like a soda left out overnight. All the fizz has left. That doesn't mean that the ellipse has no flavor! It has its own unique parts and equations, depending on whether (you guessed it) it's horizontal or vertical. Here's what you need to know about any ellipse, whether it's horizontal or vertical:

- The center is at the point (h, v).

- The longer axis of symmetry is called the *major axis,* and the distance from the center to a point on the ellipse along the major axis is represented by a. The points where the ellipse intersects this axis are called the *vertices.*

- The shorter axis of symmetry is called the *minor axis,* and the distance from the center to a point on the ellipse along the minor axis is represented by b. The points where the ellipse intersects this axis are called the *co-vertices.*

- This means that a is always greater than b in an ellipse.

- You can find the foci of the ellipse along its major axis by using the equation $f^2 = a^2 - b^2$.

Figure 12-5 shows how all these pieces fall into place for a horizontal and a vertical ellipse.

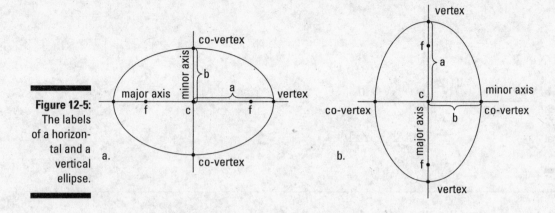

Figure 12-5:
The labels
of a horizon-
tal and a
vertical
ellipse.

Short and fat: The horizontal ellipse

The equation of a horizontal ellipse is:

$$\frac{\left(x-h\right)^2}{a^2} + \frac{\left(y-v\right)^2}{b^2} = 1$$

Notice that all the variables (except for f, the focus) we mention make their appearance in the equation. Notice also that because $a > b$, for any ellipse, $a^2 > b^2$. The fact that the bigger number, a^2, is in the denominator of the x fraction tells you that the ellipse is horizontal. To graph any horizontal ellipse after it's written in this form, mark the center first. Then count out a units to the left and right and b units up and down. These four points determine the ellipse's shape. The vertices are points found at $(h \pm a, v)$. The co-vertices are $(h, v \pm b)$. The two foci are f units in the same direction as a. As points, the foci are $(h \pm f, v)$.

Q. State the center, vertices, and foci of the ellipse $3x^2 - 18x + 5y^2 = 3$.

A. **Center: (3, 0); vertices: $(3 \pm \sqrt{10}, 0)$; foci: (5, 0) and (1, 0).** As usual, you need to complete the square to write this ellipse in its proper form. The constant is already on the opposite side, so begin by factoring: $3(x^2 - 6x) + 5y^2 = 3$. Next, complete the square and balance the equation: $3(x^2 - 6x + 9) + 5y^2 = 3 + 27$. Factor the perfect square: $3(x - 3)^2 + 5y^2 = 30$. Divide everything by 30: $\frac{\left(x-3\right)^2}{10} + \frac{y^2}{6} = 1$. This tells you the center is (3, 0). Then, if $a^2 = 10$, $a = \pm\sqrt{10}$, and if $b^2 = 6$, $b = \pm\sqrt{6}$. This gives you the vertices at $(3 \pm \sqrt{10}, 0)$. It also tells you the co-vertices are at $(3, \pm\sqrt{6})$. Lastly, $f^2 = 10 - 6$, so $f^2 = 4$, which tells you that $f = \pm 2$ and the foci are at (5, 0) and (1, 0).

Q. Sketch the graph of the ellipse $\frac{\left(x+2\right)^2}{25} + \frac{\left(y-1\right)^2}{16} = 1$.

A. **See Figure 12-6.** This ellipse is written in the proper form, so to graph it, all you have to do is identify its parts. The center is (–2, 1). If $a^2 = 25$, then $a = \pm 5$. This means your vertices are 5 units to the left and the right from the center, at (–7, 1) and (3, 1). If $b^2 = 16$, then $b = \pm 4$. This means your co-vertices are 4 units above and below the center at (–2, 5) and (–2, –3). This graph is shown in Figure 12-6.

Figure 12-6:
Graphing a
horizontal
ellipse.

In figure: V (−7, 1), C (−2, 1), V (3, 1), $\frac{(x+2)^2}{25} + \frac{(y-1)^2}{16} = 1$

7. Sketch the graph of the ellipse $4x^2 + 12y^2 - 8x - 24y = 0$.

Solve It

8. Write the equation of the ellipse with vertices at $(-1, 1)$ and $(9, 1)$ and foci at $(4 \pm \sqrt{21}, 1)$.

Solve It

Tall and skinny: The vertical ellipse

The equation of a vertical ellipse is:

$$\frac{(x-h)^2}{b^2} + \frac{(y-v)^2}{a^2} = 1$$

This equation looks awfully familiar, doesn't it? The only difference between a horizontal ellipse and a vertical one is the location of a. When the bigger number is under x, it's a horizontal ellipse. When the bigger number is under y, it's a vertical ellipse. You graph this ellipse by marking the center, counting up and down a units to find the vertices, and then counting left and right b units to find the co-vertices. This means your vertices are at $(h, v \pm a)$ and your co-vertices are at $(h \pm b, v)$. Your foci move in the same direction as your vertices, so they're at $(h, v \pm f)$.

Q. Sketch the graph of the ellipse $81x^2 + 4y^2 = 324$. State the foci of this ellipse.

A. **See Figure 12-7; the foci are $(0, \pm \sqrt{77})$.** We wanted to include an example like this because most of our students freeze when presented with a question like this one. Just remember that your only goal is to write the equation in the proper form. The equation has no x variable or y variable to the first degree. That means you don't have to complete the square! Say what? All you

need to do is get 1 on the right side of the equation by dividing everything by 324. When you do, it reduces to $\dfrac{x^2}{4} + \dfrac{y^2}{81} = 1$, which also conveniently puts it in the form you want. This ellipse has its center at the origin $(0, 0)$. Then, $a^2 = 81$, so a moves up and down 9 units, while $b^2 = 4$, so b moves left and right 2 units. This gives you the graph in Figure 12-7. Lastly, $f^2 = 81 - 4 = 77$, so $f = \pm\sqrt{77}$. This means your foci are at $(0, \pm\sqrt{77})$.

81x^2 + 4y^2 = 324

Figure 12-7:
Graphing a
vertical
ellipse.

Q. Write the equation of the vertical ellipse with its center at $(-4, 1)$ if its major axis has a length of 10 and its minor axis has a length of 8.

A. $\dfrac{(x+4)^2}{16} + \dfrac{(y-1)^2}{25} = 1$. You have all the information you need to write the equation. The center is given to you as $(-4, 1)$. If the major axis has a length of 10, then $2a = 10$, or $a = 5$. Also, the minor axis has a length of 8, so $2b = 8$, or $b = 4$. Knowing that the ellipse is vertical tells you to put a^2 under y and b^2 under x. This gives you the equation $\dfrac{(x+4)^2}{16} + \dfrac{(y-1)^2}{25} = 1$.

9.   Sketch the graph of the ellipse $\dfrac{(x-1)^2}{8} + \dfrac{(y+2)^2}{6} = 1.$

Solve It

10.   State the ordered pair for the vertices, the co-vertices, and the foci of the ellipse $(x+1)^2 + \dfrac{(y-4)^2}{16} = 1.$

Solve It

Graphing Hyperbolas: No Caffeine Required

A *hyperbola* is the set of all points where the difference in the distance from two fixed points (the foci) to any point is constant. Hyperbolas always come in pairs. Each one is a perfect mirror reflection of the other. Maybe they're the narcissists of the math world, always checking themselves out in the mirror that is their axis of symmetry. There are horizontal and vertical hyperbolas. Regardless of how the hyperbola opens, you always find the following parts:

✔ The center is at the point (h, v).

✔ The graph on both sides gets closer and closer to two diagonal lines known as *asymptotes*. The equation of the hyperbola, regardless of whether it's horizontal or vertical, gives you two values: a and b. These help you draw a box, and when you draw the diagonals of this box, you find the asymptotes.

✔ There are two axes of symmetry:

   • The one passing through the *vertices* is called the *transverse axis*. The distance from the center along the transverse axis to the vertex is represented by a.

   • The one perpendicular to the transverse axis through the center is called the *conjugate axis*. The distance along the conjugate axis from the center to the edge of the box that determines the asymptotes is represented by b.

   • a and b have no relationship; a can be less than, greater than, or equal to b.

✔ You can find the foci by using the equation $f^2 = a^2 + b^2$.

Figure 12-8 shows the parts of a hyperbola — one horizontal and one vertical.

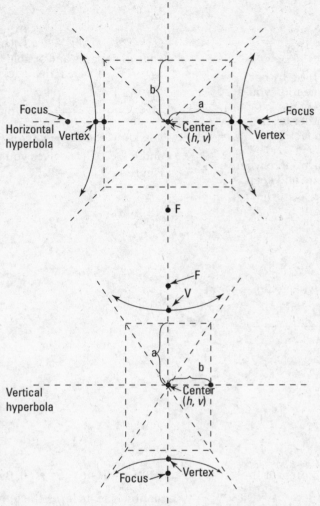

Figure 12-8:
Horizontal
and vertical
hyperbolas
and their
parts.

Hip horizontal hyperbolas (alliteration!)

The horizontal hyperbola is *so* hip (how hip is it?) that it doesn't even have to stand up — it just lies around all day. The equation of a horizontal hyperbola is:

$$\frac{(x-h)^2}{a^2} - \frac{(y-v)^2}{b^2} = 1$$

This one looks really similar to the equation of the horizontal ellipse. But if you look closely, you notice the subtraction sign in between the two fractions. To begin graphing, identify a, which helps determine one edge of a box that you can use to find the hyperbola's asymptotes. The corners of this imaginary box are two points of the asymptotes, so they can be used to draw those lines. The value of a is in the denominator of the x fraction, so it will be left and right from the center. The vertices are at $(h \pm a, v)$. The other edge of the box is found from b, under the y fraction. It moves up and down. The foci move in the same direction as a and can be found at $(h \pm f, v)$. The equation of the asymptotes of a horizontal hyperbola is given by $y = \pm \frac{b}{a}(x-h)+v$.

Q. Sketch the graph of $2x^2 - 3y^2 + 10x + 6y = 4\frac{1}{2}$.

A. **See Figure 12-9.** Put it in its form by completing the square. (Did we mention you'd be experts at this by the end of this chapter?) Write the equation in order first: $2x^2 + 10x - 3y^2 + 6y = 4\frac{1}{2}$. Then factor out the coefficients: $2(x^2 + 5x) - 3(y^2 - 2y) = 4\frac{1}{2}$. Watch out for the negative sign there when factoring. Complete the square and keep the equation balanced: $2(x^2 + 5x + \frac{25}{4}) -$

$3(y^2 - 2y + 1) = 4\frac{1}{2} + \frac{25}{2} - 3$. Factor the difference of squares and simplify: $2(x + \frac{5}{2})^2 - 3(y - 1)^2 = 30$. Last, divide everything by 30: $\dfrac{\left(x + \frac{5}{2}\right)^2}{15} - \dfrac{\left(y - 1\right)^2}{10} = 1$. From the equation you know that the center is $(-\frac{5}{2}, 1)$, $a = \sqrt{15}$, and $b = \sqrt{10}$. This gives you the graph in Figure 12-9.

$$2x^2 - 3y^2 + 10x + 6y = \frac{41}{2}$$

Figure 12-9:
Graphing a horizontal hyperbola.

Q. Find the equation of the asymptotes for the hyperbola $\dfrac{x^2}{25} - \dfrac{\left(y - 6\right)^2}{9} = 1$.

A. $y = \pm\dfrac{5}{3}(x - 0) + 6 = \pm\dfrac{5}{3}x + 6$. Because this equation is in its form, the information is easy to find. $h = 0$, $v = 6$, $a = 5$, and $b = 3$. Put those into the equation to find the asymptotes for a horizontal hyperbola and get $y = \pm\dfrac{5}{3}(x - 0) + 6 = \pm\dfrac{5}{3}x + 6$.

11. Find the equation of the hyperbola that has its center at (4, 1), its vertex at (7, 1), and one of its asymptotes is $3y = 2x - 5$.

Solve It

12. Sketch the graph of the equation $(2x - y)(x + 5y) - 9xy = 10$.

Solve It

Vexing vertical vyperbolas (er, hyperbolas)

The equation of a vertical hyperbola is:

$$\frac{(y - v)^2}{a^2} - \frac{(x - h)^2}{b^2} = 1$$

Do you see the differences between the horizontal and vertical hyperbolas? The x and y switch places (along with the h and v). The a stays on the left, and the b stays on the right. When you write a hyperbola in its form, you need to make sure that the positive squared term is always first. The vertices are at $(h, v \pm a)$ and the foci are at $(h, v \pm f)$.

You can find the asymptotes using the equation $y = \pm \frac{a}{b}(x - h) + v$, sort of like the horizontal hyperbolas.

Q. Sketch the graph of the hyperbola
$16y^2 - 25x^2 + 64y - 50x = 361$.

A. **See Figure 12-10.** Start by completing the
square. Rewrite the equation: $16y^2 + 64y - 25x^2 - 50x = 361$. Factor out the coefficients:
$16(y^2 + 4y) - 25(x^2 + 2x) = 361$. Now, complete
the square (keep the equation balanced,
too): $16(y^2 + 4y + 4) - 25(x^2 + 2x + 1) = 361 + 64 - 25$. Factor, factor: $16(y + 2)^2 - 25(x + 1)^2 = 400$. Divide the 400:

$\dfrac{(y+2)^2}{25} - \dfrac{(x+1)^2}{16} = 1$. The center of this

hyperbola is $(-1,-2)$, where $a = 5$ (and goes
up and down) and $b = 4$ (left/right). This
finally gives you Figure 12-10.

Figure 12-10:
Graphing a
vertical
hyperbola.

Q. Two science stations are two miles apart. They record an explosion, one station two seconds after
the other. Write the equation of the hyperbola that describes the situation by placing both stations on the y-axis with the origin at the center. (**Note:** Sound travels at 1100 feet per second.)

A. $\dfrac{y^2}{1210000} - \dfrac{x^2}{26668400} = 1$. If the two stations are two miles apart, that means one is a mile below the

origin $(0, -1)$ and the other is a mile above it $(0, 1)$. Right away, you should notice that all other
measurements in this problem are given in feet, so you have to convert 1 mile to 5,280 feet. This
means the microphones of the two stations are at $(0, -5{,}280)$ and $(0, 5{,}280)$. This gives you the foci
of the hyperbola: $f = 5{,}280$. To find the vertices, you can set up two different equations based on
the distance $(d = rt)$ the explosion travels to get to the two stations: $d_1 = 1100t$ and $d_2 = 1100(t + 2)$.
The absolute value of the difference of the distances gives you the vertices of the hyperbola:
$\left| d_1 - d_2 \right| = \left| 1100t - 1100(t + 2) \right| = \left| 1100t - 1100t - 2200 \right| = \left| -2200 \right| = 2200$. a for your vertex is half of this:
$^{2200}\!/\!_2 = 1100$, so $a^2 = 1{,}210{,}000$. Now that you know f and a, you can find b^2 from the equation
$f^2 = a^2 + b^2$: $b^2 = 26{,}668{,}400$. Lastly, because the hyperbola has its center at the origin, you can write

the equation of the hyperbola as $\dfrac{y^2}{1210000} - \dfrac{x^2}{26668400} = 1$.

13. Sketch the graph of the equation $x^2 + 2x - 4y^2 + 32y = 59$.

Solve It

14. Write the equation of the hyperbola that has its center at $(-3, 5)$, one vertex at $(-3, 1)$, and passes through the point $(1, 5 - 4\sqrt{2})$.

Solve It

Identifying Conic Sections

Often, you'll be presented with an equation and asked to graph it, but you won't be told what type of conic section it is. You have to be able to identify what type it is before doing any work, but that's easier than it sounds, because there are only the four conics, and they have distinct differences:

- **Circles** have x^2 and y^2 with equal coefficients on both.
- **Parabolas** have x^2 or y^2, but not both.
- **Ellipses** have x^2 and y^2 with different (not equal) coefficients on each.
- **Hyperbolas** have x^2 and y^2 where exactly one coefficient is negative.

Table 12-1 has all the information you need to know about the four conics in one handy-dandy chart.

Table 12-1	Types of Conic Sections and Their Parts
Type of Conic	*Parts of the Conic*
Circle	Center (h, v)
	Radius r
Horizontal parabola	Vertex (h, v)
	Focus $(h + \frac{1}{4a}, v)$
	Directrix $x = h - \frac{1}{4a}$
	Axis of symmetry $x = h$
Vertical parabola	Vertex (h, v)
	Focus $(h, v + \frac{1}{4a})$
	Directrix $y = v - \frac{1}{4a}$
	Axis of symmetry $y = v$
Horizontal ellipse	Center (h, v)
	Vertices $(h \pm a, v)$
	Co-vertices $(h, v \pm b)$
	Foci $(h \pm f, v)$
Vertical ellipse	Center (h, v)
	Vertices $(h, v \pm a)$
	Co-vertices $(h \pm b, v)$
	Foci $(h, v \pm f)$
Horizontal hyperbola	Center (h, v)
	Vertices $(h \pm a, v)$
	Foci $(h \pm f, v)$
	Asymptotes $y = \pm \frac{b}{a}(x - h) + v$
Vertical hyperbola	Center (h, v)
	Vertices $(h, v \pm a)$
	Foci $(h, v \pm f)$
	Asymptotes $y = \pm \frac{a}{b}(x - h) + v$

15. Sketch the graph of $3x^2 + 4y^2 - 6x + 16y - 5 = 0$.

Solve It

16. Sketch the graph of $4x^2 - 8x - 1 = 4y^2 - 4y$.

Solve It

17. Sketch the graph of $4(x - 2) = 2y^2 + 6y$.

Solve It

18. Sketch the graph of $2y^2 - 4x^2 + 8x - 8 = 0$.

Solve It

19. Sketch the graph of $4x^2 + 4y^2 - 8y + 16x - 4 = 0$.

Solve It

20. Sketch the graph of $3x^2 - 4y^2 + 3x - 2y - \frac{25}{2} = 0$.

Solve It

Converting from Parametric Form to Polar Coordinates and Back

So far, you've graphed all the conics in *rectangular form* (x, y). However, you can graph a conic section in two other ways:

- ✔ **Parametric form:** This form is for conics that can't be easily written as a function $y = f(x)$. Both x and y are written in two different equations as being dependent on one other variable (usually t).

- ✔ **Polar form:** You recognize this from Chapter 11, where every point is expressed as (r, θ).

We show you how to deal with both of these forms in the following sections.

Parametric form for conic sections

Parametric form defines both x and y in terms of another arbitrary value called the *parameter*. Most often, this is represented by t, as real-world applications set the definitions based on time. You can find x and y by picking values for t. Why change? In parametric form you can find how far an object has moved over time (the x equation) and the object's height over time (the y equation).

Q. Sketch the curve given by the parametric equations $x = 2t + 1$, $y = t^2 - 3t + 1$, and $1 < t \leq 5$.

A. **See Figure 12-11.** Create a table for t, x, and y. Pick values of t between the interval values given to you, and then figure out what the x and y values are for each t value. Table 12-2 shows these values, and Figure 12-11 shows the graph of this parametric function.

Even though $t > 1$ in the given interval, you need to start your table off with this value to see what the function would have been. Your graph has an open circle on it at this point to indicate that the value isn't included in the graph or the interval.

Table 12-2	Plug and Chug a Parametric Equation	
t value	*x value*	*y value*
1	3	−1
2	5	−1
3	7	1
4	9	5
5	11	11

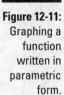

Figure 12-11:
Graphing a function written in parametric form.

21. Sketch the graph of the parametric equations $x = t^2 - 1$, $y = 2t$, and $-2 \le t \le 3$.

Solve It

22. Sketch the graph of the parametric equations $x = \dfrac{1}{\sqrt{t-1}}$, $y = \dfrac{1}{t-1}$, and $t > 1$.

Solve It

Changing from parametric form to rectangular form

The only other way to graph a parametric curve is to write it in rectangular form. To do this, you must solve one equation for the parameter and then substitute that value into the other equation. It's easiest if you pick the equation you *can* solve for the parameter (choose the equation that's linear). To show you how it works, we use the example from the last section.

Q. Write the parametric equations $x = 2t + 1$, $y = t^2 - 3t + 1$, and $1 < t \le 5$ in rectangular form.

A. $y = \dfrac{1}{4}x^2 - 2x + \dfrac{11}{4}$. First, solve the equation that's linear for t: $t = \dfrac{x-1}{2}$. Then substitute this value into the other equation for t: $y = \left(\dfrac{x-1}{2}\right)^2 - 3\left(\dfrac{x-1}{2}\right) + 1$. Simplify this equation to get $y = \dfrac{1}{4}x^2 - 2x + \dfrac{11}{4}$.

23. Eliminate the parameter and find an equation in x and y whose graph contains the curve of the parametric equations $x = t^2$, $y = 1 - t$, and $t \geq 0$.

Solve It

24. Eliminate the parameter of the parametric equations $x = t - 5$ and $y = \sqrt{t}$.

Solve It

Conic sections on the polar coordinate plane

Conic sections on the polar coordinate plane are all based on a special value known as *eccentricity,* or *e.* This value describes what kind of conic section it is, as well as the conic's shape. It's difficult to know what kind of conic section you're dealing with until you know what the eccentricity is:

- If $e = 0$, the conic is a circle.
- If $0 < e < 1$, the conic is an ellipse.
- If $e = 1$, the conic is a parabola.
- If $e > 1$, the conic is a hyperbola.

When you know e, all conics are expressed in polar form based on (r, θ), where r is the radius and θ is the angle. See Chapter 11 for more information on polar equations.

All conics in polar form are written based on four different equations:

$$r = \frac{ke}{1 - e\cos\theta} \quad \text{or} \quad \frac{ke}{1 - \sin\theta}$$

$$r = \frac{ke}{1 + e\cos\theta} \quad \text{or} \quad \frac{ke}{1 + \sin\theta}$$

where e is eccentricity and k is a constant value. To graph any conic section in polar form, substitute values of θ and plug and chug away until you get a picture!

Q. Graph the equation $r = \dfrac{2}{4 - \cos\theta}$

A. **See Figure 12-12.** First, notice that the equation as shown doesn't fit exactly into any of the equations we just gave you. All those denominators begin with 1, and this equation begins with 4! To deal with this, factor out the 4 from the denominator to

get $\dfrac{2}{4\left(1 - \dfrac{1}{4}\cos\theta\right)}$, which is the same as

$\dfrac{2 \cdot \dfrac{1}{4}}{1 - \dfrac{1}{4}\cos\theta}$. Notice that this makes e the

same in the numerator and denominator ($\frac{1}{4}$) and that k is 2. Now that you know e is $\frac{1}{4}$, that tells you the equation is an ellipse. Plugging in values gives you points $\theta = 0$, $r = \frac{2}{3}$; $\theta = \frac{\pi}{2}$, $r = \frac{1}{2}$; $\theta = \pi$, $r = \frac{2}{5}$; $\theta = \frac{3\pi}{2}$, $r = \frac{1}{2}$. This gives you the ellipse in Figure 12-12.

Figure 12-12: Graphing a conic section written in polar form.

25. Graph the equation of $r = \dfrac{8}{1 + \cos\theta}$ and label any vertices.

Solve It

26. Identify the conic section whose equation is $r = \dfrac{12}{3 - 4\sin\theta}$ by stating its eccentricity.

Solve It

Answers to Problems on Conic Sections

1 Find the center and radius of the circle $2x^2 + 2y^2 - 4x = 15$. Then graph the circle. **The center is (1, 0) and the radius is ≈ 2.92.**

Rewrite the equation so the x and y variables are together to get $2x^2 - 4x + 2y^2 = 15$. Factor out the coefficient: $2(x^2 - 2x) + 2y^2 = 15$. Complete the square: $2(x^2 - 2x + 1) + 2y^2 = 15 + 2$. Factor and get $2(x - 1)^2 + 2y^2 = 17$. Divide everything by 2 to write the circle in its form: $(x - 1)^2 + y^2 = 8.5$. This means the center is (1, 0) and the radius is $\sqrt{8.5}$, or about 2.92.

2 Write the equation of the circle with the center (–1, 4) if the circle passes through the point (3, 1). **The answer is $(x + 1)^2 + (y - 4)^2 = 25$.**

If you're given the center and a point, you can find the distance between the two points using the distance formula from Chapter 1:

$$d = \sqrt{\left(x_2 - x_1\right)^2 + \left(y_2 - y_1\right)^2} = \sqrt{\left(3 + 1\right)^2 + \left(1 - 4\right)^2} = \sqrt{4^2 + 3^2} = \sqrt{16 + 9} = \sqrt{25} = 5.$$

This tells you the radius of the circle. Now that you know both the radius and the center, you can write the equation: $(x + 1)^2 + (y - 4)^2 = 25$.

3 What's the vertex of the parabola $y = -x^2 + 4x - 6$? Sketch the graph of this parabola. **The answer is (2, –2); see the following graph.**

Add the 6 to both sides: $y + 6 = -x^2 + 4x$. Now, factor out the coefficient: $y + 6 = -1(x^2 - 4x)$. Complete the square and balance the equation: $y + 6 - 4 = -1(x^2 - 4x + 4)$. Simplify and factor: $y + 2 = -1(x - 2)^2$. Lastly, subtract 2 from both sides to write the equation in its proper form: $y = -1(x - 2)^2 - 2$. This means the vertex is located at the point (2, –2).

4 Find the focus and the directrix of the parabola $y = 4x^2$. **The focus is (0, ¹⁄₁₆) and the directrix is $y = -¹⁄₁₆$.**

There's no square to complete, so if it helps you to fill in the missing information with zeros, then rewrite the equation as $y = 4(x - 0)^2 + 0$. This puts the vertex at the origin (0, 0). Because $a = 4$, the focus is $\frac{1}{4a}$ units above this point at (0, ¹⁄₁₆) and the directrix is the line that runs $\frac{1}{4a}$ units below the vertex, perpendicular to the axis of symmetry at $y = -¹⁄₁₆$.

5 Sketch the graph of $x = 2(y - 4)^2$. **See the graph for the answer.**

This equation is written in the proper form, unless you'd like to rewrite it as $x = 2(y - 4)^2 + 0$ because the h is missing. This is a horizontal parabola with its vertex at (0, 4). It opens to the right, with a horizontal transformation of 2.

6 Determine whether this parabola opens left or right: $x = -y^2 - 7y + 3$. **The answer is the parabola opens to the left.**

Did we get you on this one? Did you start completing the square? We know you're used to doing it by now, but make sure you look at the directions of the problem. This one just asks you whether the parabola opens to the left or right, and you can tell from the leading coefficient of –1 (without doing any work at all) that the parabola opens to the left. If you did actually want to graph this equation, the standard form is $y = -1(x + \frac{1}{2})^2 + \frac{6}{4}$.

7 Sketch the graph of the ellipse $4x^2 + 12y^2 - 8x - 24y = 0$. **See the graph for the answer.**

You have to complete the square twice for this one, so maybe we made up for that last question here. There's no constant to move, so rewrite the equation and factor out the coefficients: $4(x^2 + 2x) + 12(y^2 - 2y) = 0$. Now complete the square and balance the equation: $4(x^2 + 2x + 1) + 12(y^2 - 2y + 1) = 0 + 4 + 12$. Factor: $4(x + 1)^2 + 12(y - 1)^2 = 16$. Divide everything by 16:

$\dfrac{(x-1)^2}{16} + \dfrac{3(y-1)^2}{4} = 1$. But don't start to graph it yet, as each ellipse has a coefficient of 1 on

the numerator, and you have to divide that 3 in the second fraction to get $\dfrac{(x-1)^2}{16} + \dfrac{(y-1)^2}{4/3} = 1$.

Now that the equation is written in standard form, you can graph it. The center is (1, 1), the vertices are (3, 1) and (–1, 1), and the co-vertices are (1, 2.2) and (1, –0.2).

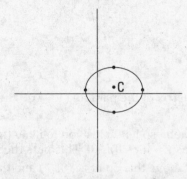

8 Write the equation of the ellipse with vertices at (–1, 1) and (9, 1) and foci at $(4 \pm \sqrt{21}, 1)$. **The answer is $\dfrac{(x-4)^2}{25} + \dfrac{(y+1)^2}{4} = 1$.**

Knowing the vertices tells you the center, because it's halfway between them (the midpoint of the segment connecting them — see Chapter 1 for a refresher). This means the center is at (4, 1) and that each vertex is 5 units away from the center, so $a = 5$ and $a^2 = 25$. The foci are $\pm\sqrt{21}$

units away from the center, which tells you that $f = \sqrt{21}$. Now that you know a and f, you can

find b^2 using $f^2 = a^2 - b^2$. In this case, $b^2 = 4$. Now you can write the equation: $\dfrac{(x-4)^2}{25} + \dfrac{(y+1)^2}{4} = 1$.

9 Sketch the graph of the ellipse $\dfrac{(x-1)^2}{8} + \dfrac{(y+2)^2}{6} = 1$. **See the graph for the answer.**

How convenient! This equation is written in the proper form, so you don't have to complete the square. Did we make up for our sneaky question in number 6, yet? The center is (1, –2), $a \approx 2.83$, and $b \approx 2.45$. That gives you the following ellipse:

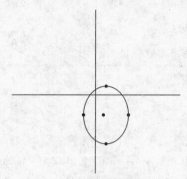

10 State the ordered pair for the vertices, the co-vertices, and the foci of the ellipse

$(x+1)^2 + \dfrac{(y-4)^2}{16} = 1$. **The vertices are (–1, 8) and (–1, 0); the co-vertices are (–2, 4) and (0, 4); and the foci are $(-1, 4 \pm \sqrt{15})$.**

The sneakiest part of this problem is that the x half of the equation isn't written as a fraction,

but that's easy to remedy by writing the denominator of 1: $\dfrac{(x+1)^2}{1} + \dfrac{(y-4)^2}{16} = 1$. This tells you

that the center is (–1, 4), $a = 4$, and $b = 1$. The vertices are $(-1, 4 \pm 4) = (-1, 8)$ and (–1, 0). The

co-vertices are $(-1 \pm 1, 4) = (-2, 4)$ and (0, 4). Lastly, $f^2 = a^2 - b^2$, so $f = \pm \sqrt{15}$, which gives you

the foci at $(-1, 4 \pm \sqrt{15})$.

11 Find the equation of the hyperbola that has its center at (4, 1), its vertex at (7, 1), and one of its

asymptotes is $3y = 2x - 5$. **The answer is $\dfrac{(x-4)^2}{9} - \dfrac{(y-1)^2}{4} = 1$.**

You're given the center (4, 1) and the equation of the asymptote, which you can rewrite in

slope-intercept form by dividing by 3 to get $y = \dfrac{2}{3}x - \dfrac{5}{3}$. Because the vertex is 3 units to the

right, this is a horizontal hyperbola. The slope of the asymptote, $\dfrac{2}{3}$, is the value of $\dfrac{b}{a}$. If we lost

you there, you have to write the equation of the asymptote in point-slope form because you know

the point is the center (4, 1) and the slope is $m = \frac{2}{3}$. This gives you the equation $y - 1 = \dfrac{2}{3}(x-4)$.

Adding 1 to both sides makes the equation look like the equation of the asymptote for a hori-

zontal hyperbola and helps you identify $\dfrac{b}{a}$ at the same time. Now that you know the center, a,

and b, you can write the equation: $\dfrac{(x-4)^2}{9} - \dfrac{(y-1)^2}{4} = 1$.

12 Sketch the graph of the equation $(2x - y)(x + 5y) - 9xy = 10$. **See the graph for the answer.**

Does the term $-9xy$ make you a little nervous? Don't be. Just FOIL out the binomials and get $2x^2 + 10xy - xy - 5y^2 - 9xy = 10$. Notice that all the xy terms cancel to give you $2x^2 - 5y^2 = 10$. You can divide everything by 10 to write this equation in its form and get $\dfrac{x^2}{5} - \dfrac{y^2}{2} = 1$. The center of this hyperbola is at $(0, 0)$; $a = \sqrt{5}$, or ≈ 2.24; and $b = \sqrt{2}$, or ≈ 1.41.

13 Sketch the graph of the equation $x^2 + 2x - 4y^2 + 32y = 59$. **See the graph for the answer.**

You just knew we were gonna say, "Complete the square," didn't you? Okay, let's get going. Factor out the coefficients, including the 1 in front of the x^2: $1(x^2 + 2x) - 4(y^2 - 8y) = 59$. Complete the square and balance away: $1(x^2 + 2x + 1) - 4(y^2 - 8y + 16) = 59 + 1 - 64$. Factor and simplify: $1(x + 1)^2 - 4(y - 4)^2 = -4$. Divide everything by 4: $\dfrac{(x+1)^2}{-4} + (y - 4)^2 = 1$. Did you notice how it suddenly became a vertical hyperbola because the y fraction is positive? You have to rewrite it to put it in its correct form: $(y - 4)^2 - \dfrac{(x+1)^2}{4} = 1$. That gives you the following figure.

14 Write the equation of the hyperbola that has its center at $(-3, 5)$, one vertex at $(-3, 1)$, and passes through the point $(1, 5 - 4\sqrt{2})$. **The answer is** $\dfrac{(y-5)^2}{16} - \dfrac{(x+3)^2}{16} = 1.$

This one requires a thinking cap, that's for sure. We suggest drawing it out on a sheet of graph paper and marking the center, the vertex, and the point, but that's it. Doing so shows you that it's a vertical hyperbola, so start with the form for *any* vertical hyperbola: $\dfrac{(y-v)^2}{a^2} - \dfrac{(x-h)^2}{b^2} = 1.$

Based on what you're given, you know that $h = -3$, $v = 5$, and the vertex is 4 units below the center, so $a = 4$. What do you do with the point the question gives you? Remember that all points are (x, y), so $x = 1$ and $y = 5 - 4\sqrt{2}$. Plug all these values into the equation that you started with and get $\dfrac{\left(5-4\sqrt{2}-5\right)^{2}}{4^{2}}-\dfrac{\left(1--3\right)^{2}}{b^{2}}=1$, which means you only have one variable to solve for,

b^{2}. This equation simplifies to $2-\dfrac{16}{b^{2}}=1$. Solving it gets you $b^{2}=16$. This means you can finally

write the equation: $\dfrac{\left(y-5\right)^{2}}{16}-\dfrac{\left(x+3\right)^{2}}{16}=1$.

15 Sketch the graph of $3x^2 + 4y^2 - 6x + 16y - 5 = 0$. **See the graph for the answer.**

This is an ellipse because x and y are both squared but have different coefficients. Add 5 to both sides: $3x^2 + 4y^2 - 6x + 16y = 5$. Rewrite the equation with the x and the y together: $3x^2 - 6x + 4y^2 + 16y = 5$. Factor the coefficients: $3(x^2 - 2x) + 4(y^2 + 4y) = 5$. Complete the square and balance the equation: $3(x^2 - 2x + 1) + 4(y^2 + 4y + 4) = 5 + 3 + 16$. Factor the perfect square trinomials and simplify: $3(x - 1)^2 + 4(y + 2)^2 = 24$. Divide everything by 24: $\dfrac{\left(x-1\right)^{2}}{8}+\dfrac{\left(y+2\right)^{2}}{6}=1$. That gives you the following graph.

16 Sketch the graph of $4x^2 - 8x - 1 = 4y^2 - 4y$. **See the graph for the answer.**

Rewrite the equation first: $4x^2 - 8x - 4y^2 + 4y = 1$. You should recognize that you have a hyperbola on your hands because there are an x^2 and a y^2 where exactly one has a negative coefficient. Factor out the coefficients: $4(x^2 - 2x) - 4(y^2 - y) = 1$. Complete the square and balance the equation: $4(x^2 - 2x + 1) - 4(y^2 - y + \frac{1}{4}) = 1 + 4 - 1$. Factor the perfect squares: $4(x - 1)^2 - 4(y - \frac{1}{2})^2 = 4$. Divide everything by 4: $\left(x-1\right)^{2}-\left(y-\dfrac{1}{2}\right)^{2}=1$. In this particular hyperbola, the values of a and b are both 1. Knowing this and the center gives you the following graph.

17 Sketch the graph of $4(x - 2) = 2y^2 + 6y$. **See the graph for the answer.**

Notice right away that the equation doesn't have an x^2, so this is a horizontal parabola. Go ahead and distribute the 4 first: $4x - 8 = 2y^2 + 6y$. Now, factor the coefficient on the y^2 variable: $4x - 8 = 2(y^2 + 3y)$. Completing the square for this one gets you fractions; half of 3 is $\frac{3}{2}$ and that value squared is $\frac{9}{4}$. Add this inside the parentheses, and don't forget to add $2 \cdot \frac{9}{4}$ to the other side to keep the equation balanced: $4x - 8 + \frac{9}{2} = 2(y^2 + 3y + \frac{9}{4})$. Simplify and factor: $4x - \frac{7}{2} =$

$2(y + \frac{3}{2})^2$. Begin to solve for x by adding over the $\frac{7}{2}$: $4x = 2(y + \frac{3}{2})^2$. Now divide each term by 4: $x = \frac{1}{2}\left(y + \frac{3}{2}\right)^2 + \frac{7}{8}$. This gives you a horizontal parabola with vertex at $(\frac{7}{8}, -\frac{3}{2})$.

18 Sketch the graph of $2y^2 - 4x^2 + 8x - 8 = 0$. **See the graph for the answer.**

You have a hyperbola to graph this time. We'll skip the narrative on how to complete the square and show the steps only (we're confident you're a pro at it by now): $2y^2 - 4x^2 + 8x = 8$; $2y^2 - 4(x^2 - 2x) = 8$; $2y^2 - 4(x^2 - 2x + 1) = 8 - 4$; $2y^2 - 4(x - 1)^2 = 4$; $\frac{y^2}{2} - (x - 1)^2 = 1$. This gives you the following hyperbola.

19 Sketch the graph of the equation $4x^2 + 4y^2 - 8y + 16x - 4 = 0$. **See the graph for the answer.**

You should recognize that this is a circle because of the x^2 and y^2 with equal coefficients on both. Here are the steps to completing the square: $4x^2 + 4y^2 - 8y + 16x = 4$; $4x^2 + 16x + 4y^2 - 8y = 4$; $4(x^2 + 4x) + 4(y^2 - 2y) = 4$; $4(x^2 + 4x + 4) + 4(y^2 - 2y + 1) = 4 + 16 + 4$; $4(x + 2)^2 + 4(y - 1)^2 = 24$. Because this is a circle, you need to get coefficients of 1 in front of both sets of parentheses by dividing by 4: $(x + 2)^2 + (y - 1)^2 = 6$. This circle has its center at $(-2, 1)$ and its radius is $\sqrt{6}$.

20 Sketch the graph of $3x^2 - 4y^2 + 3x - 2y - \frac{25}{2} = 0$. **See the graph for the answer.**

This is another hyperbola because the coefficient on y^2 is negative while the coefficient on x^2 is positive. Here are the usual steps: $3x^2 + 3x - 4y^2 - 2y = \frac{25}{2}$; $3(x^2 + x) - 4(y^2 + \frac{1}{2}y) = \frac{25}{2}$;

$3(x^2 + x + \frac{1}{4}) - 4(y^2 + \frac{1}{2}y + \frac{1}{16}) = \frac{25}{2} + \frac{3}{4} - \frac{1}{4}$; $3(x + \frac{1}{2})^2 - 4(y + \frac{1}{4})^2 = 12$; $\dfrac{\left(x + \frac{1}{2}\right)^2}{4} - \dfrac{\left(y + \frac{1}{4}\right)^2}{3} = 1$.

This gives the following hyperbola.

21 Sketch the graph of the parametric equations $x = t^2 - 1$, $y = 2t$, and $-2 \leq t \leq 3$. **See the graph for the answer.**

Set up a table of t, x, and y where you pick the t and find x and y. Be sure to stay within the interval defined by the problem. Here's our chart:

t	x	y
-2	3	-4
-1	0	-2
0	-1	0
1	0	2
2	3	4
3	8	6

These (x, y) points give you the following graph:

22 Sketch the graph of the parametric equations $x = \dfrac{1}{\sqrt{t-1}}$, $y = \dfrac{1}{t-1}$, and $t > 1$. **See the graph for the answer.**

Another table comes in handy here:

t	x	y
1	undefined	undefined
2	1	1
3	$\dfrac{1}{\sqrt{2}}$	½
4	$\dfrac{1}{\sqrt{3}}$	⅓
5	½	¼

These points give you the following graph:

23 Eliminate the parameter and find an equation in x and y whose graph contains the curve of the parametric equations $x = t^2$, $y = 1 - t$, and $t \geq 0$. **The answer is $y = 1 \pm \sqrt{x}$.**

Solve for t in the first equation by taking the square root of both sides: $\pm \sqrt{x} = t$. Substitute this value into the other equation: $y = 1 \pm \sqrt{x}$.

Some teachers may require that you don't leave the square root in this equation and instead write it in a form that looks more like a conic section. Subtract 1 from both sides: $y - 1 = \pm \sqrt{x}$. Then square both sides to get $(y - 1)^2 = x$. That looks more like a parabola now, doesn't it?

24 Eliminate the parameter of the parametric equations $x = t - 5$ and $y = \sqrt{t}$. **The answer is $y = \sqrt{x + 5}$.**

The first equation is easy to solve for t by adding 5 to both sides: $x + 5 = t$. Substitute this value into the other equation and get $y = \sqrt{x + 5}$.

If you're required by your teacher to write this equation without the square root, square both sides to get $y^2 = x + 5$. This, too, looks like a parabola.

25 Graph the equation of $r = \dfrac{8}{1 + \cos\theta}$ and label any vertices. **See the graph for the answer.**

Notice that the given equation is the same thing as $\dfrac{8 \cdot 1}{1 + 1\cos\theta}$, which tells you that $k = 8$ and $e = 1$. This makes it a parabola.

26 Identify the conic section whose equation is $r = \dfrac{12}{3 - 4\sin\theta}$ by stating its eccentricity. **The answer is this conic is a hyperbola because** $e = \frac{4}{3}$.

You have to factor the 3 out of the denominator first: $\dfrac{12}{3\left(1 - \dfrac{4}{3}\sin\theta\right)} = \dfrac{4}{1 - \dfrac{4}{3}\sin\theta}$. This is the same

thing as $\dfrac{3 \cdot \dfrac{4}{3}}{1 - \dfrac{4}{3}\sin\theta}$, and that tells you that $k = 3$ and $e = \frac{4}{3}$, which is why this one is a hyperbola.

Chapter 13

Finding Solutions for Systems of Equations

....

....

No, a *system of equations* is *not* how to organize, arrange, or classify them. A system of equations is a collection of linear equations involving the same set of variables. The point is to find one solution, if there is one, that works in all the equations. Solving one equation for one variable is almost always possible and usually pretty easy to do. Two variables require at least two equations, and three variables require at least three. You see, you need a unique equation for every variable present if you have to solve the system.

It goes without saying that the bigger the system of equations becomes, the longer and harder it may be to solve. Solving a system involves several techniques, and sometimes it may be easier to solve certain systems certain ways. That's why math textbooks show all the techniques — so you know when to use each technique.

Of course, you can choose to always solve all systems using one technique, but another technique may require fewer steps, which will save you time, not to mention money (money for aspirin, that is, for all the headaches you'd get if you solved all equations one specific way!).

A Quick-and-Dirty Technique Overview

Here's a handy guide to all the techniques we cover in this chapter and when it's best to use them:

✔ If a system has two or three variables, you can use *substitution* or *elimination* to solve.

✔ If a system has four or more variables, you should use *matrices*, in which you have the following choices:

- The Gaussian method
- Inverse matrices
- Cramer's Rule

We discuss all these techniques in detail in this chapter. And if that's not enough information for you, we also discuss *systems of inequalities,* which can only be solved by graphing.

Solving Two Linear Equations with Two Variables

When you're presented with a system of linear equations in two variables, the best methods to solve them are known as *substitution* and *elimination.* As we mention earlier in this chapter, you can use either method to solve any system of this type, but textbooks usually show you both methods because each one has its unique advantages. In keeping with this spirit, we show you both methods, but we also explain *when* to use both as well.

Just remember that with each system of equations in two variables, you have to find the solution to *both* variables, usually x and y. Don't stop until you have both, or else you may only get half the credit on your next test for doing half the work.

Also remember that sometimes, systems of equations don't have a solution. Remember when you used to graph two straight lines on one graph to determine the point of intersection? Well, because that's the least accurate of all the methods to solve, mathematicians came up with the other methods that we talk about in this chapter. However, if you recall, sometimes the two lines were parallel to each other — without an intersection! This meant that there was *no solution.* The fact that there may be no solution may pop up from time to time using these other methods as well.

So how will you recognize a system of equations with no solution without using a graph? That's easy — you end up with an equation that just doesn't make sense. It may say 2 = 7 or –1 = 10, but you'll know right away that it has no solution. It's also possible that you're given the same line (in disguise) twice. If you were to graph that system, you'd end up with one line on top of another.

These two lines share infinite points, so you say that the system has *infinite solutions.* These equations boil down at some point to an *identity* — the left and right sides of the equation are exactly the same (such as 2 = 2, or $10x = 10x$, or $4y - 3 = 4y - 3$), and these, too, are easy to recognize.

The substitution method

In the *substitution method,* you solve one equation for one variable and then substitute this expression for that variable in the other equation. If one of the two equations you're given has already been solved for one variable, huge bells and whistles should go off inside your head. You know you've got a winner for the substitution method. Of course, if one equation can be easily solved for one variable (one variable has a coefficient of 1), you also know that substitution is a good bet.

EXAMPLE

Q. Solve the system of equations:

$$\begin{cases} x = 4y - 1 \\ 2x + 5y = 11 \end{cases}$$

A. **x = 3, y = 1.** Notice how the first equation says "x = . . ."? This tells you to use substitution. You can take this expression and substitute it into the other equation where it says x. This gets you $2(4y - 1) + 5y = 11$. The substitution method makes your job easier because you end up with one equation in one variable — and this one is easy to solve! When you do, you get $y = 1$. Now

that you know half of your answer (y in this case), you can substitute that value into one of the original equations to get the other half (x).

TIP

Save yourself some time and steps by substituting the first answer you get into the equation that has already been solved for a variable. For this example, because you know that $x = 4y - 1$ and you figure out that $y = 1$, it takes very few steps to figure out that $x = 3$.

1. Use substitution to solve the system

$$\begin{cases} r + s = 6 \\ s = 13 - 2r \end{cases}$$

Solve It

2. The sum of two numbers is 14 and their difference is 2. Find the numbers.

Solve It

The elimination method

PRE-CALC RULES

$$\frac{\begin{array}{r} 1 \\ +1 \end{array}}{2}$$

Elimination is the method of choice when both of the linear equations given to you are written in *standard form:*

$Ax + By = C$

$Dx + Ey = F$

where A, B, C, D, E, and F are all real numbers. It's called *standard form* because it's supposed to be *the* standard way that textbooks depict linear equations. But the truth is that textbook authors like to keep you on your toes so they write linear equations in all kinds of forms.

In the two equations, if the x and y terms are opposite of each other, then you should choose elimination. In the *elimination method,* you add the two equations together so that one of the variables disappears (is eliminated). Sometimes, however, you must multiply one or both equations by a constant in order for the terms to have opposite signs. This way, when you add the two equations together, one of the variables will be eliminated.

Q. Solve the system $\begin{cases} 2x - y = 6 \\ 3x + y = 4 \end{cases}$.

A. **$x = 2$, $y = -2$.** Notice that you *could* solve this system using substitution, because in the second equation, you can solve for y by subtracting $3x$ from both sides. But notice that the y terms are exact opposites of each other. If you add the two equations you get $5x = 10$. This means that you can divide both sides by 5 to easily solve for x and get $x = 2$. Substituting this back into either equation, you find that $y = -2$.

Q. Solve the system $\begin{cases} 2x - 3y = 5 \\ 4x + 5y = -1 \end{cases}$.

A. **$x = 1$, $y = -1$.** To solve this system by substitution is possible, but it would eventually mean dividing one of the coefficients and creating those ugly fractions we all hate. Instead, you can avoid the fractions by using the elimination method. The fact that both equations are written in standard form is another vote in favor of the elimination method. Notice that the y terms have opposite signs, so you can eliminate them (you can eliminate any variable you choose, but it's all about the least amount of steps). It's a little like you're finding the least common multiple of both coefficients, in this case the 3 and the 5. The smallest number that both of those go into is 15, so you have to multiply the top equation by 5 and the bottom equation by 3. This gives you $\begin{cases} 10x - 15y = 25 \\ 12x + 15y = -3 \end{cases}$. Adding these two equations together gives you $22x = 22$, which gives you the solution $x = 1$. You then have to substitute this value back into one of the two original equations to solve for y. In this example, $y = -1$.

3. Solve the system $\begin{cases} \dfrac{x}{2} - \dfrac{y}{3} = -3 \\ \dfrac{2x}{3} + \dfrac{y}{2} = -2 \end{cases}$.

Solve It

4. Solve the system $\begin{cases} 3x - 2y = 4 \\ 6x - 4y = 8 \end{cases}$.

Solve It

Not-So-Straight: Solving Nonlinear Systems

The substitution and elimination methods are common tools for systems of equations that include *nonlinear equations*. Yes, now at least one of your two given equations will often be a quadratic equation (it could also be a rational function or some other type). The method you choose to use for these types of systems depends on the types of equations that you're given. We break the following sections into those types and show you how to best solve each one.

One linear equation and one not

When one equation is linear and the other equation isn't linear, it's best to use the substitution method. That's because the linear equation can be easily solved for one variable. You can then substitute this value into the other equation to solve. Most often, that means solving a quadratic equation at some point, so if you need to brush up on those techniques, see Chapter 4.

Q. Solve the system of equations $\begin{cases} x^2 + y = 0 \\ 2x - y = 3 \end{cases}$.

A. $x = -3, y = -9; x = 1, y = -1$. As we mention earlier, it's usually easier to solve the linear equation first. The second given equation is the linear one, and it's easier to solve for y (no pesky coefficients to divide). Doing so gets you $y = 2x - 3$. After you substitute this expression into the first equation for y,

you get $x^2 + 2x - 3 = 0$. This quadratic polynomial factors to $(x + 3)(x - 1) = 0$. Then, using the zero product property (for more information, see Chapter 4), you get two solutions: $x = -3$ or $x = 1$. Uh-oh. Now what? No stopping you! Two solutions for x means twice the substitution and twice the y answers. If $x = -3$, then $y = -9$, and if $x = 1$, then $y = -1$.

---

5. Solve $\begin{cases} x^2 - y = 1 \\ x + y = 5 \end{cases}$.

Solve It

6. Solve the system of equations $\begin{cases} x + y = 9 \\ xy = 20 \end{cases}$.

Solve It

Two equations that are nonlinear

In this section, we explore what happens when both of the given equations are nonlinear. These steps will also, most likely, create one final quadratic equation to solve. Solving these systems requires the elimination method because one of the quadratic terms must cancel to solve for the other variable. This is probably easiest to see with an example, so read on.

Q. Solve the system $\begin{cases} x^2 + y^2 = 25 \\ x^2 - y = 5 \end{cases}$

A. **$(x, y) = (0, -5), (\pm 3, 4)$.** Notice right away that the x^2 terms in both equations have the same coefficient. If you multiply the second equation by -1 and then add the two equations together, you get $y^2 + y = 20$. Subtracting the 20 from both sides gets you the quadratic equation, $y^2 + y - 20 = 0$, that you have to factor: $(y + 5)(y - 4) = 0$. Solve and get $y = -5$ and $y = 4$. Substituting $y = -5$ into the second equation gets you $x^2 + 5 = 5$, or $x^2 = 0$, which means that $x = 0$. Substituting $y = 4$ into the same equation gets you $x^2 - 4 = 5$ or $x^2 = 9$, which gives you $x = \pm 3$. Both of these solutions work.

7. Solve $\begin{cases} x^2 + y^2 = 1 \\ x + y^2 = -5 \end{cases}$

Solve It

8. Solve $\begin{cases} 27x^2 - 16y^2 = -400 \\ -9x^2 + 4y^2 = 36 \end{cases}$

Solve It

Systems of equations disguised as rational equations

Sometimes you'll see two equations that look like they're rational equations. That means that the variable is in the denominator of the equation. As we discuss in Chapter 3, sometimes rational functions have undefined values. Keep this in mind with the final solutions you find to the equation — they may not really work! *Always* check the solutions to these types of equations because you never know which ones are actually solutions and which ones aren't until you double-check. We recommend that you always start off by substituting the rational expressions into the other equation so you can deal with a more normal-looking system (as if there is such a thing).

EXAMPLE

Q. Solve the system $\begin{cases} \dfrac{3}{x} + \dfrac{2}{y} = 11 \\ \dfrac{1}{x} - \dfrac{1}{y} = -11 \end{cases}$

A. $x = -\frac{5}{11}$, $y = \frac{5}{44}$. Okay, so you see all those fractions right off the bat and you decide to throw in the towel and walk away, right? Wrong! Make the fractions go away by starting off with a substitution. Notice, first of all, that you can rewrite the given

system as $\begin{cases} 3\dfrac{1}{x} + 2\dfrac{1}{y} = 11 \\ \dfrac{1}{x} - \dfrac{1}{y} = -11 \end{cases}$. By letting $u = \dfrac{1}{x}$

and $v = \dfrac{1}{y}$, you can conveniently rewrite the

entire system as $\begin{cases} 3u + 2v = 11 \\ u - v = -11 \end{cases}$. You can then

use any of the methods you're already familiar with to solve the system. For this example, if you multiply the second equation by 2, you get $2u - 2v = -22$. You can then add the two equations to eliminate v and get $5u = -11$. This means that $u = -\frac{11}{5}$. Then you can work your way backwards to get $x = -\frac{5}{11}$. We'd then suggest substituting your u value to find your v, which gets you on your way to finding that $y = \frac{5}{44}$.

9. Solve $\begin{cases} \dfrac{14}{x+3} + \dfrac{7}{4-y} = 9 \\ \dfrac{21}{x+3} - \dfrac{3}{4-y} = 0 \end{cases}$

Solve It

10. Solve $\begin{cases} \dfrac{12}{x+1} - \dfrac{12}{y-1} = 8 \\ \dfrac{6}{x+1} - \dfrac{6}{y-1} = -2 \end{cases}$

Solve It

Solving More than Two (Equations and/or Variables)

Now that you have a grip on solving a system of equations in two variables, we'll change gears to systems with three variables. Remember that for each variable you need at least one unique equation. In other words, two variables need at least two equations, three variables need at least three equations, four variables need four, and so on. Why change to three variables, though? Well, we live in a three-dimensional world, so we need three variables to represent it.

TIP

Most of the time when you're given systems larger than 2×2, you want to use elimination. You have to take two equations at a time and eliminate one variable. Then, you have to take another two equations and eliminate the same variable. If you start with a 3×3 system, this will knock you down to a 2×2 system, which you then must solve. If you start with a 4×4 system, you work down to a 3×3 system, which you must then work down to 2×2 and solve. Sounds fun, doesn't it?

Q. Solve the system of equations
$$\begin{cases} x - y + z = 6 \\ x - z = -2 \\ x + y = -1 \end{cases}.$$

A. **x = 1, y = –2, z = 3.** To make things easier in this section in general, we'll always label the given equations from top to bottom with capital letters. For this example, the top equation $x - y + z = 6$ is equation A, the middle equation is equation B and the bottom equation is equation C. Notice that equation C is already missing the z variable. Also notice that equations A and B have z variables that are exact opposites of each other.

If you add those two together (A + B), you'd get a brand new equation (we'll call it D) which is $2x - y = 4$.

If you then take equation D and add C to it (D + C), you'd get one more equation, E, which is $3x = 3$, which is solved easily to get $x = 1$. You can then substitute this value into equation C and get $1 + y = -1$, which means that $y = -2$. You can also use the fact that $x = 1$ in equation B to get $1 - z = -2$, or $-z = -3$, or finally $z = 3$.

Q. Solve the system
$$\begin{cases} x + y + z + w = -1 \\ 2x + y + z = 0 \\ 2y + z - w = -6 \\ x - z + 2w = 7 \end{cases}.$$

A. **x = 2, y = –1, z = –3, and w = 1.** Most of our students panic with systems that are bigger than 3×3, so we decided to show one example of a 4×4 so you can see how it works. A: $x + y + z + w = -1$; B: $2x + y + z = 0$; C: $2y + z - w = -6$; D: $x - z + 2w = 7$. Notice that B is missing its w variable, so if you can use the others to eliminate w as well, that'll be the first success.

A + C gives you a new equation, E: $x + 3y + 2z = -7$.

–2A + D gives a new equation, F: $-x - 2y - 3z = 9$.

So now you have three equations with three variables: B, D, and F. Next, you have to pick another variable to eliminate — it doesn't matter which one; we eliminate x.

E + F gives you equation G: $y - z = 2$.

Meanwhile, B + 2F also eliminates x with equation H: $-3y - 5z = 18$.

Finally, 3G + H eliminates y with one last equation, J: $-8z = 24$; $z = -3$.

Using this solution in equation H gives you $-3y - 5(-3) = 18$; $-3y + 15 = 18$; $-3y = 3$; $y = -1$. Now that you know z and y, use them in equation E: $x + 3(-1) + 2(-3) = -7$; $x - 3 - 6 = -7$; $x = 2$. Then use all those answers in equation A: $2 - 1 - 3 + w = -1$; $w = 1$.

11. Solve the system $\begin{cases} 3x - 2y = 17 \\ x - 2z = 1 \\ 3y + 2z = 1 \end{cases}$

Solve It

12. Solve $\begin{cases} 2x - y + z = 1 \\ x + y - z = 2 \\ -x - y + z = 2 \end{cases}$

Solve It

13. Solve $\begin{cases} 2x + 3y + 4z = 37 \\ 4x - 3y + 2z = 17 \\ x + 2y - 3z = -5 \\ 3x - 2y + z = 11 \end{cases}$

Solve It

14. Solve $\begin{cases} 3a + b + c + d = 0 \\ 4a + 5b + 2c = 15 \\ 4a + 2b + 5d = -10 \\ -5a + 3b - d = 8 \end{cases}$

Solve It

Graphing Systems of Inequalities

A *system of inequalities* is more than one inequality in more than one variable. Up until now in your math career, you've probably seen systems of inequalities that have all been linear. In pre-calc, you continue with those types of problems but then move up to nonlinear systems of inequalities. That's right, you'll be seeing quadratics and conics, too. For a review of how to graph *one* inequality, see Chapter 1. If you're quacky on quadratics, see Chapter 3. And if conics sound crazy, see Chapter 12. If, on the other hand, you're raring to go, read on.

The only way to solve a system of inequalities is to graph it. You end up with (hopefully) two overlapping shaded regions — the overlap is the solution. Every single point in the overlap is a solution to the system. What happens if there's no overlap? Well, there's no solution!

When you multiply or divide an inequality by a negative number, the inequality sign changes: < becomes >, ≤ becomes ≥, and vice versa. This is a pretty important fact to remember because it affects your shading in the end.

Q. Sketch the graph of the inequality
$\begin{cases} 3x + y \leq 5 \\ x + 2y \leq 4 \end{cases}$

A. **See Figure 13-1.** Because both of these inequalities are linear, you have to put them in slope-intercept form to graph

them. The top equation in slope-intercept form is $y \leq -3x + 5$, and the bottom

equation becomes $y \leq -\frac{1}{2}x + 2$. Graphing

both on the same coordinate plane gives you Figure 13-1.

Figure 13-1:
Graphing a
system of
linear
inequalities.

$$\begin{cases} y \le \dfrac{-1x}{2} + 2 \\ y \le -3x + 5 \end{cases}$$

EXAMPLE

Q. Sketch the graph of the inequality
$$\begin{cases} x^2 + y^2 < 16 \\ x - 2y > -4 \end{cases}$$

A. **See Figure 13-2.** This time, the top equation is a circle and the bottom equation is a line. The circle is in the proper form to graph, so you don't have to do any work there (other than to graph, that is), while the bottom equation in slope-intercept

form is $y < \dfrac{1}{2}x + 2$. For these types of problems, we recommend that you pick test points to see where to shade. For example, the origin $(0, 0)$ is a great point to try in the original equations to see whether it works. Is $0^2 + 0^2 < 16$? Yes, so you shade inside the circle. In the second equation, is $0 - 2(0) > -4$? Yes, so you also shade below the line. See the graph in Figure 13-2.

Figure 13-2:
Another
graph of a
system of
inequalities;
this time,
one equa-
tion isn't
linear.

$$\begin{cases} x^2 + y^2 < 16 \\ y < \dfrac{1x}{2} + 2 \end{cases}$$

15. Sketch the graph of $\begin{cases} 2x + y \ge 9 \\ 2x - y \ge 1. \\ x \le 7 \end{cases}$

Solve It

16. Sketch the graph of $\begin{cases} x^2 + y^2 \ge 9 \\ x^2 + (y-3)^2 \ge 9 \end{cases}$

Solve It

17. Sketch the graph of $\begin{cases} x^2 - y > 2 \\ x - y < 4 \end{cases}$.

Solve It

18. Sketch the graph of $\begin{cases} x \geq 0 \\ x + y < 4 \\ y \leq \sqrt{x-1} \end{cases}$.

Solve It

We're Partial to . . . Decomposing Partial Fractions!

Call us crazy, but we're partial to partial fractions! (We know, but we really couldn't resist.) The process known as *decomposing partial fractions* takes one fraction and breaks it down as the sum or difference of two (or more) other fractions. This requires being an expert at factoring, so if you need a review, turn to Chapter 4 and read up on how to do it. If you're a pro by now, then you know to always follow these general steps:

1. **Factor the denominator.**

2. **Write separate fractions, one for each factor of the denominator based on these rules:**

   a. If the factor is linear, it has some constant in the numerator.

   b. If the factor is quadratic, it has a linear expression in the numerator.

Note: If any factor has a power on it, you have to create one fraction for each power, from 1 on up to the highest degree. This is probably best shown with an example. Suppose that you're able to factor the polynomial in a particularly long denominator into $(x-4)(x+1)^3(2x-1)(3x^2-6x+2)^2$. You would have to create the sum of seven different fractions: $\dfrac{A}{x-4} + \dfrac{B}{x+1} + \dfrac{C}{(x+1)^2} + \dfrac{D}{(x+1)^3} + \dfrac{E}{2x-1} + \dfrac{Fx+G}{3x^2-6x+2} + \dfrac{Hx+J}{(3x^2-6x+2)^2}$.

The first denominator's factor is linear, so its numerator is a constant. The second denominator's factor is linear with a degree of 3, so you need to create three different constant numerators: one for the first degree, the second for the second degree, and the third for the third degree. The third factor is also linear, so it got one fraction with a constant on the top. Lastly, the final term is quadratic and second degree, so it got two linear terms in the numerator: one for the first degree and the second for the second degree.

Q. Find the partial fraction decomposition of $\dfrac{7x+5}{x^2+x-2}$.

A. $\dfrac{4}{x-1}+\dfrac{3}{x+2}$. Begin by factoring the denominator of the given fraction into $\dfrac{7x+5}{x^2+x-2}=\dfrac{7x+5}{(x+2)(x-1)}$.

Because both factors in the denominator are linear, you break them into two different fractions with constant numerators: $\dfrac{7x+5}{(x-1)(x+2)}=\dfrac{A}{x-1}+\dfrac{B}{x+2}$. Now, multiply every fraction by the factored denominator, which results in some big-time cancelling:

$\dfrac{7x+5}{\cancel{(x-1)}\cancel{(x+2)}}\cancel{(x-1)}\cancel{(x+2)}=\dfrac{A}{\cancel{x-1}}\cancel{(x-1)}(x+2)+\dfrac{B}{\cancel{x+2}}(x-1)\cancel{(x+2)}$. You now have the

simplified equation $7x+5=A(x+2)+B(x-1)$. When you multiply this out, you get the equation $7x+5=Ax+2A+Bx-B$. Now, gather like terms: $7x+5=Ax+Bx+2A-B$. Factor the x out on the right side: $7x+5=(A+B)x+2A-B$. Notice how both sides match up, which means that

$7x=(A+B)x$, or $7=A+B$ *and* $5=2A-B$. This gives you a system of equations to solve: $\begin{cases}7=A+B\\5=2A-B\end{cases}$.

Now you know why the textbooks usually include this material in the chapter with systems of equations (as well as why we include it here — aren't we clever?). Add these two equations to eliminate B and get $12=3A$, or $4=A$. Substituting this into the top equation gets you $7=4+B$, or $3=B$. You can now use these values to write the sum of two fractions: $\dfrac{4}{x-1}+\dfrac{3}{x+2}$.

---

19. Find the constants A and B:
$$\dfrac{x-38}{x^2+x-12}=\dfrac{A}{x+4}+\dfrac{B}{x-3}.$$

Solve It

20. Find the form of the partial fraction decomposition for $\dfrac{5x-4}{(x-1)^2}$, but don't find the constants.

Solve It

---

21. Find the partial fraction decomposition for
$$\dfrac{2x^2-21x+18}{(x-1)(x^2-4x+4)}.$$

Solve It

22. Find the partial fraction decomposition for
$$\dfrac{11x^2-7x+14}{2x^3-4x^2+3x-6}.$$

Solve It

There Is No Spoon: Working with a Matrix

A *matrix* is a collection of numbers arranged in rows and columns. Each number inside the matrix is called an *element*. A matrix comes in handy when you have a bunch of data that you need to keep track of. Usually, a matrix is named by its *dimensions,* or how big it is. This is also sometimes known as the *order* of the matrix and is always the number of rows by the number of columns. For example, if matrix M is 4×3, it has four rows and three columns. After the data is organized in this fashion, you can add, subtract, and even multiply matrices.

There's no such thing as matrix division, so don't worry about it. There's also an operation known as *scalar multiplication*, which means you multiply the entire matrix by a constant.

To add or subtract matrices, you have to operate on their corresponding elements. In other words, you add or subtract the first row/first column in one matrix to or from the exact same element in another matrix. The two matrices must have the same dimensions; otherwise, an element in one matrix won't have a corresponding element in the other.

Figure 13-3 shows two matrices and what their sum and differences are. Figure 13-4 shows the scalar multiplication 3A.

$$A = \begin{bmatrix} -5 & 1 & -3 \\ 6 & 0 & 2 \\ 2 & 6 & 1 \end{bmatrix} \qquad B = \begin{bmatrix} 2 & 4 & 5 \\ -8 & 10 & 3 \\ -2 & -3 & -9 \end{bmatrix}$$

Figure 13-3: Addition and subtraction of matrices.

$$A + B = \begin{bmatrix} -3 & 5 & 2 \\ -2 & 10 & 5 \\ 0 & 3 & -8 \end{bmatrix} \qquad A - B = \begin{bmatrix} -7 & -3 & -8 \\ 14 & -10 & -1 \\ 4 & 9 & 10 \end{bmatrix}$$

Figure 13-4: Multiplying matrix A by 3.

$$3A = 3 \begin{bmatrix} -5 & 1 & -3 \\ 6 & 0 & 2 \\ 2 & 6 & 1 \end{bmatrix} = \begin{bmatrix} -15 & 3 & -9 \\ 18 & 0 & 6 \\ 6 & 18 & 3 \end{bmatrix}$$

Multiplying matrices is another can of worms. First of all, to multiply two matrices AB (the matrices are written right next to each other, with no symbol in between), the number of columns in matrix A *must* match the number of rows in matrix B. If matrix A is $m \times n$ and matrix B is $n \times p$, the product AB has dimensions $m \times p$. And remember, when it comes to matrix multiplication, AB doesn't equal BA; in fact, just because AB exists doesn't even mean that BA does as well.

For all problems in this section, $M = \begin{bmatrix} -5 & -1 & 3 & 6 \\ 0 & 2 & -2 & 6 \end{bmatrix}$, $N = \begin{bmatrix} 2 & 4 & 5 & -8 \\ 10 & 3 & -2 & -3 \end{bmatrix}$, and

$$P = \begin{bmatrix} -1 & 2 & -1 \\ 4 & 4 & 0 \\ 2 & 3 & 1 \\ -5 & 2 & -1 \end{bmatrix}.$$

Q. Find 3M – 2N.

A. $\begin{bmatrix} -19 & -11 & -1 & 34 \\ -20 & 0 & -2 & 24 \end{bmatrix}$.

First, substitute in each given equation:

$3\begin{bmatrix} -5 & -1 & 3 & 6 \\ 0 & 2 & -2 & 6 \end{bmatrix} - 2\begin{bmatrix} 2 & 4 & 5 & -8 \\ 10 & 3 & -2 & -3 \end{bmatrix}$.

Follow the order of operations and multiply in the scalars:

$\begin{bmatrix} -15 & -3 & 9 & 18 \\ 0 & 6 & -6 & 18 \end{bmatrix} - \begin{bmatrix} 4 & 8 & 10 & -16 \\ 20 & 6 & -4 & -6 \end{bmatrix}$.

Then subtract the two matrices, watching the negative signs: $\begin{bmatrix} -19 & -11 & -1 & 34 \\ -20 & 0 & -2 & 24 \end{bmatrix}$.

Q. Find MP.

A. $\begin{bmatrix} -23 & 7 & 2 \\ -26 & 14 & -8 \end{bmatrix}$

You need to multiply each element of each row of the left matrix by each element of each column of the right matrix.

The sum of the first row times the first column: $-5(-1) - 1(4) + 3(2) + 6(-5) = -23$. This is the first row, first column answer.

The sum of the first row times the second column: $-5(2) - 1(4) + 3(3) + 6(2) = 7$. This is the first row, second column answer.

The sum of the first row times the third column: $-5(-1) - 1(0) + 3(1) + 6(-1) = 2$. This is the first row, third column answer.

The sum of the second row times the first column: $0(-1) + 2(4) - 2(2) + 6(-5) = -26$. This is the second row, first column answer.

The sum of the second row times the second column: $0(2) + 2(4) - 2(3) + 6(2) = 14$. This is the second row, second column answer.

The sum of the second row times the third column: $0(-1) + 2(0) - 2(1) + 6(-1) = -8$. This is the second row, third column answer.

Putting these all into a matrix gives you the answer $\begin{bmatrix} -23 & 7 & 2 \\ -26 & 14 & -8 \end{bmatrix}$.

23. Find 4N.

Solve It

24. Find 4N + 5M.

Solve It

25. Find 3M – P.

Solve It

26. Find NP.

Solve It

Getting It in the Right Form: Simplifying Matrices

You can write any system of equations in *matrix form*. To do so, follow these steps:

1. **Write all the coefficients in one matrix, called the *coefficient matrix.*** Each equation gets its own row in the matrix, and each variable gets its own column, written in the same order as the equations.

2. **Multiply this times another column matrix with all the variables in it, called the *variable matrix,*** in order from top to bottom.

3. **Set this product equal to a column matrix with the answers in it, sometimes called the *answer matrix.***

Row echelon form is similar to matrix form; however, you only work with the coefficient matrix. Across any row, the first number element (besides 0) that you run into is called the *leading coefficient.* For a coefficient matrix to be in row echelon form,

✔ Any row with all 0s in it must be the bottom row.

✔ The leading coefficient in any row must be to the right of the leading coefficient in the row above it.

Reduced row echelon form takes row echelon form and makes all the leading coefficients the number 1. Also, each element above or below a leading coefficient must be 0. Figure 13-5 shows a matrix (a) in reduced row echelon form and a matrix (b) not in reduced row echelon form.

Figure 13-5:
A matrix (a) in reduced row echelon form and (b) not in reduced row echelon form.

a.   b.

Finally, *augmented form* takes the coefficient matrix and tacks on an extra column — a column with the answers in it so that you can look at the entire system in one convenient package.

These ways of writing systems of equations in matrices come in handy when dealing with systems that are 4×4 or larger. Your goal is to get the matrix into row echelon form using *elementary row operations.* These operations are different from the operations we discuss in the previous section because they're done on only one row at a time. Here are three row operations you can perform:

✔ Multiply each element of a row by a constant

✔ Interchange any two rows

✔ Add two rows together

We stay consistent with *Pre-Calculus For Dummies* and use the same notation that we use there to represent these elementary row operations. This means that $4r_2 \rightarrow r_2$ multiplies the second row by 4 to change the second row. $r_1 \leftrightarrow r_3$ swaps row one with row three. $r_3 + r_1 \rightarrow r_1$ adds row three to row one and changes row one. $4r_2 + r_1 \rightarrow r_1$ first multiplies row two by 4 and then adds that to row one to change row one.

You can use any combination of these row operations to get the given matrix into row echelon form. Use reduced row echelon form only if you're specifically told to do so by your teacher or textbook, as it takes more steps.

We only focus on the forms and the row operations in this section. To really dig in deep and discover how to get a matrix in row echelon form, read on to the next section.

EXAMPLE

Q. Write the system of equations $\begin{cases} 3x - y = 6 \\ 2x + 3y = 3 \end{cases}$ as a matrix system.

A. $\begin{bmatrix} 3 & -1 \\ 2 & 3 \end{bmatrix}\begin{bmatrix} x \\ y \end{bmatrix} = \begin{bmatrix} 6 \\ 3 \end{bmatrix}$. The matrix on the left is the coefficient matrix, containing all the coefficients from the system. The second matrix is the variable matrix, and the third one, on the right, is the answer matrix. This completes the job for this question.

Q. Write the system from the previous question as an augmented matrix.

A. $\begin{bmatrix} 3 & -1 & | & 6 \\ 2 & 3 & | & 3 \end{bmatrix}$. Just take the coefficient matrix and add on the answer matrix. *Voilà!* You've got yourself an augmented matrix. Notice that the vertical line separates the two and lets you know that this is an augmented matrix and not a normal 2×3 matrix.

---

27. Using the augmented matrix from the last example, use elementary row operations to find $-3r_2 \rightarrow r_2$.

Solve It

28. Now, using your answer from 27, find $r_1 \leftrightarrow r_2$.

Solve It

---

29. Now, keep going and find $r_1 + r_2 \rightarrow r_2$.

Solve It

30. Lastly, find $3r_2 + r_1 \rightarrow r_1$.

Solve It

Solving Systems of Equations Using Matrices

You can solve a system of equations using matrices in three ways. Putting a matrix in row echelon form using the techniques described in the last section is called *Gaussian elimination.* The second way uses a method called *inverse matrices,* and the third method is called *Cramer's Rule.* Your book may cover all these techniques, or it may cover only one, or it may not even cover this material at all. But we're here for you if you need us! We dedicate one section to each of these ways, to keep it nice and simple.

Gaussian elimination

The process of putting a matrix in row echelon form is called *Gaussian elimination*. We focus in this section on matrices in augmented form because that's most commonly what you'll be asked to do, but know that the rules don't change if you're asked to do this with some other form of matrix equation. The goals of using the elementary row operations are simple: get a 1 in the upper-left corner of the matrix, get 0s in all positions underneath this 1, get 1s for all leading coefficients diagonally from the upper-left to the lower-right corners, and then get 0s below each of them. When you get to that point, you use a process called *back substitution* to solve for all the variables in the system.

Q. Put the system of equations $\begin{cases} 3x - y = 6 \\ 2x + 3y = 3 \end{cases}$ in augmented form and then write the equation in row echelon form.

A. $\begin{bmatrix} 1 & -1/3 & | & 2 \\ 0 & 11/3 & | & -1 \end{bmatrix}$, $x = {}^{21}/_{11}$, $y = {}^{-3}/_{11}$. You wrote this system as an augmented matrix $\begin{bmatrix} 3 & -1 & | & 6 \\ 2 & 3 & | & 3 \end{bmatrix}$ in the last section. Now you need to get it into row echelon form. First, you need to get a 0 below the element in the upper-left corner. This is easiest if you get a 1 in the upper-left corner first. In fact, some textbooks may say that row echelon form has leading coefficients of

1 for this reason. Follow the elementary row operation $\frac{1}{3}r_1 \to r_1$ to get $\begin{bmatrix} 1 & -1/3 & | & 2 \\ 2 & 3 & | & 3 \end{bmatrix}$. Now, to make the first element of row two 0, you need to add -2. So take $-2r_1 + r_2 \to r_2$ to get $\begin{bmatrix} 1 & -1/3 & | & 2 \\ 0 & 11/3 & | & -1 \end{bmatrix}$. This gives you an equation in the second row that's easy to solve for y: $\frac{11}{3}y = -1$, or $y = {}^{-3}/_{11}$. Now you can work backwards using back substitution in the top equation: $x - \frac{1}{3}y = 2$. You know the value of y so substitute that in and get $x - \frac{1}{3}\left(\frac{-3}{11}\right) = 2$, or $x = {}^{21}/_{11}$.

---

31. Solve the system of equations $\begin{cases} 2x + 5y = 7 \\ 3x - 5y = 2 \end{cases}$ by writing it in augmented form and then putting the matrix in row echelon form.

Solve It

32. Use Gaussian elimination to solve $\begin{cases} 3x - 2y + 6z = 7 \\ x - 2y - z = -2 \\ -3x + 10y + 11z = 18 \end{cases}$.

Solve It

Inverse matrices

Another way to solve a system is by using an *inverse matrix*. This process is based on the idea that if you write a system in matrix form, you'll have the coefficient matrix multiplying the variable matrix on the left side — if only you could divide a matrix you'd have it made in the shade! Well, if you look at the simple equation $3x = 12$, you can solve it by dividing both sides by 3, $\frac{3x}{3} = \frac{12}{3}$, to get $x = 4$. This is the same thing as multiplying by its multiplicative inverse on both sides, which turns out to be the same in matrices! You have to use an inverse matrix. Remember from Chapter 3 that if $f(x)$ is a function, its inverse is denoted by $f^{-1}(x)$. This is true for matrices as well: If A is the matrix, A^{-1} is its inverse. If you have three matrices (A, B, and C) and you know that $AB = C$, then you can solve for B by multiplying the inverse matrix A^{-1} on both sides: $A^{-1}[AB] = A^{-1}C$, which simplifies to $B = A^{-1}C$.

Finding a matrix's inverse

But how do you find a matrix's inverse? Realize first that only square matrices have inverses. The number of rows must be equal to the number of columns. Even then, not every square matrix has an inverse. If the determinant (which we talk more about in the next section) of a matrix is 0, it doesn't have an inverse. The definition of a determinant involves a lot of math mumbo jumbo which, in our humble opinion, won't help you much with finding the answer. We'd rather cut to the chase and simply show you how to find the determinant.

$$\begin{bmatrix} a & b \\ c & d \end{bmatrix} ad - bc$$

When a matrix does have an inverse, you can use several ways to find it depending on how big the matrix is. If it's a 2×2, you can find it by hand using a simple formula. If it's 3×3 or bigger, you *can* find it by hand, but that doesn't mean that you *should*. Most textbooks give you the inverse matrix for these bigger matrices. You can also use a graphing calculator or the Internet to find an inverse matrix.

In the meantime, if matrix A is the 2×2 matrix $\begin{bmatrix} a & b \\ c & d \end{bmatrix}$, its inverse is found using:

$$\frac{1}{ad - bc}\begin{bmatrix} d & -b \\ -c & a \end{bmatrix}$$

Using an inverse matrix to solve a system

Now that you can find the inverse matrix, all you have to do to solve is:

1. **Write the system as a matrix equation.**

2. **Create the inverse matrix.**

3. **Multiply this inverse in front of both sides of the equation.**

4. **Cancel on the left side; multiply the matrices on the right.**

5. **Multiply the scalar.**

EXAMPLE

Q. Set up the matrix equation for the system $\begin{cases} 3x - 2y = -1 \\ x + y = 3 \end{cases}$ by using inverse matrices.

A. $x = 1, y = 2$. First set up the matrix

equation $\begin{bmatrix} 3 & -2 \\ 1 & 1 \end{bmatrix}\begin{bmatrix} x \\ y \end{bmatrix} = \begin{bmatrix} -1 \\ 3 \end{bmatrix}$. Now,

find the inverse matrix using the formula from earlier in this chapter:

$$\frac{1}{3(1) - (-2)(1)}\begin{bmatrix} 1 & 2 \\ -1 & 3 \end{bmatrix} = \frac{1}{5}\begin{bmatrix} 1 & 2 \\ -1 & 3 \end{bmatrix}.$$

Now, multiply this inverse on the left of both sides of the equation:

$$\frac{1}{5}\begin{bmatrix} 1 & 2 \\ -1 & 3 \end{bmatrix}\begin{bmatrix} 3 & -2 \\ 1 & 1 \end{bmatrix}\begin{bmatrix} x \\ y \end{bmatrix} = \frac{1}{5}\begin{bmatrix} 1 & 2 \\ -1 & 3 \end{bmatrix}\begin{bmatrix} -1 \\ 3 \end{bmatrix}.$$

To multiply a matrix by its inverse cancels everything on the left except for the variable matrix,

$$\frac{1}{5}\begin{bmatrix} 1 & 2 \\ -1 & 3 \end{bmatrix}\begin{bmatrix} 3 & -2 \\ 1 & 1 \end{bmatrix}\begin{bmatrix} x \\ y \end{bmatrix} = \frac{1}{5}\begin{bmatrix} 1 & 2 \\ -1 & 3 \end{bmatrix}\begin{bmatrix} -1 \\ 3 \end{bmatrix}.$$

That means all you have to do is multiply the matrices on the right and then multiply

the scalar: $\begin{bmatrix} x \\ y \end{bmatrix} = \frac{1}{5}\begin{bmatrix} 5 \\ 10 \end{bmatrix} = \begin{bmatrix} 1 \\ 2 \end{bmatrix}$. This gives

the solutions from top to bottom as $x = 1$ and $y = 2$.

---

33. Solve the system $\begin{cases} 4x - y = -10 \\ 2x + 3y = 16 \end{cases}$ using inverse matrices.

Solve It

34. Solve the system $\begin{cases} 4x + 3y = 17 \\ 2x - y = 11 \end{cases}$ using inverse matrices.

Solve It

Cramer's Rule

Cramer's Rule is a method based on determinants of matrices that's used to solve

systems of equations. The determinant of a 2×2 matrix $\begin{bmatrix} a & b \\ c & d \end{bmatrix}$ is $ad - bc$. The

determinant of a 3×3 matrix is found using a process called *diagonals* in some

textbooks. If A = $\begin{bmatrix} a_1 & b_1 & c_1 \\ a_2 & b_2 & c_2 \\ a_3 & b_3 & c_3 \end{bmatrix}$, then first rewrite the first two columns immediately

following the third. Draw three diagonal lines from the upper left to the lower right and three diagonal lines from the lower left to the upper right, as shown in Figure 13-6.

Figure 13-6:
How to find
a matrix's
determinant.

$$|A| = \begin{vmatrix} a_1 & b_1 & c_1 \\ a_2 & b_2 & c_2 \\ a_3 & b_3 & c_3 \end{vmatrix} \begin{matrix} a_1 & b_1 \\ a_2 & b_2 \\ a_3 & b_3 \end{matrix}$$

Then multiply down the three diagonals from left to right and up the other three. Find the sum of the numbers of the products on the top and the sum of the products on the bottom. Finally, find the difference of the top and bottom. This is the same thing as:

$$\left(a_1 b_2 c_3 + b_1 c_2 a_3 + c_1 a_2 b_3 \right) - \left(a_3 b_2 c_1 + b_3 c_2 a_1 + c_3 a_2 b_1 \right)$$

For a 2×2 system $\begin{cases} ax + by = c \\ dx + ey = f \end{cases}$:

$$x = \frac{\begin{vmatrix} c & b \\ f & e \end{vmatrix}}{\begin{vmatrix} a & b \\ d & e \end{vmatrix}} \qquad y = \frac{\begin{vmatrix} a & c \\ d & f \end{vmatrix}}{\begin{vmatrix} a & b \\ d & e \end{vmatrix}}$$

Q. Solve the system of equations
$\begin{cases} -2x + 3y = 17 \\ -5x - y = 17 \end{cases}$ using Cramer's Rule.

A. **$x = -4$, $y = 3$.** Using Cramer's Rule by substituting the coefficients and the constants,

$x = \dfrac{\begin{vmatrix} 17 & 3 \\ 17 & -1 \end{vmatrix}}{\begin{vmatrix} -2 & 3 \\ -5 & -1 \end{vmatrix}}$. Find the determinants in

the numerator and the denominator:

$\dfrac{-17 - 51}{2 + 15} = \dfrac{-68}{17} = -4$. Do the same thing

for $y = \dfrac{\begin{vmatrix} -2 & 17 \\ -5 & 17 \end{vmatrix}}{\begin{vmatrix} -2 & 3 \\ -5 & -1 \end{vmatrix}} = \dfrac{-34 + 85}{2 + 15} = \dfrac{51}{17} = 3$. This is

why if you're looking for a pain-free way of solving a system, we recommend Cramer's Rule. See how easy that was?

---

35. Find the determinant of $\begin{vmatrix} 2 & -1 & 4 \\ -3 & 4 & 6 \\ -2 & -1 & 5 \end{vmatrix}$.

Solve It

36. Use Cramer's Rule to solve
$\begin{cases} 2x - y + 4z = 7 \\ -3x + 4y + 6z = -1 \\ -2x - y + 5z = 4 \end{cases}$.

Solve It

Answers to Problems on Systems of Equations

1 Use substitution to solve the system $\begin{cases} r+s=6 \\ s=13-2r \end{cases}$. **The answer is $r = 7$ and $s = -1$.**

Substituting the fact that $s = 13 - 2r$, change the first equation to, say, $r + 13 - 2r = 6$. This simplifies to $-r + 13 = 6$, or $-r = -7$, which means that $r = 7$. Now that you know this value, you can substitute it into the second equation: $s = 13 - 2(7) = 13 - 14 = -1$. The final answer: $r = 7$, $s = -1$.

2 The sum of two numbers is 14 and their difference is 2. Find the numbers. **The answer is $x = 8$ and $y = 6$.**

First, you need to change the given words into a system of equations using variables. The sum of two numbers being 14 becomes $x + y = 14$, and their difference of 2 becomes $x - y = 2$. The first equation has an x variable with a coefficient of 1, so you can solve for it easily by subtracting y from both sides: $x = 14 - y$. Now, substitute this expression for x in the other equation: $14 - y - y = 2$. Combine like terms: $14 - 2y = 2$. Solve for y: $-2y = -12$; $y = 6$. Now that you've got that on lockdown, substitute it into the other equation to solve for x: $x = 14 - 6$; $x = 8$.

3 Solve the system $\begin{cases} \dfrac{x}{2} - \dfrac{y}{3} = -3 \\ \dfrac{2x}{3} + \dfrac{y}{2} = -2 \end{cases}$. **The answer is $(x, y) = \left(\dfrac{-78}{17}, \dfrac{36}{17} \right)$.**

We didn't do it on purpose, but sometimes the answers to these questions just aren't pretty, so don't expect 'em to be. Even though you know that the answer is going to fractionville, we still recommend that the first thing you do in any equation of this type is to get rid of the fractions by multiplying every term by the LCD. The LCD for both equations in this problem turns out to

be 6, so get multiplying: $\begin{cases} 6 \cdot \dfrac{x}{2} - 6 \cdot \dfrac{y}{3} = 6 \cdot -3 \\ 6 \cdot \dfrac{2x}{3} + 6 \cdot \dfrac{y}{2} = 6 \cdot -2 \end{cases} = \begin{cases} 3x - 2y = -18 \\ 4x + 3y = -12 \end{cases}$. Now that this looks more like all

the other systems you've been dealing with, which variable would you like to eliminate? y?

Excellent choice. Multiply the top equation by 3 and the bottom by 2: $\begin{cases} 9x - 6y = -54 \\ 8x + 6y = -24 \end{cases}$. Adding

these two equations eliminates y: $17x = -78$, which means that $x = \dfrac{-78}{17}$. Substitute this value in

to solve for y: $3\left(\dfrac{-78}{17} \right) - 2y = -18$ or $\dfrac{-234}{17} - 2y = -18$; $-2y = \dfrac{-72}{17}$; $y = \dfrac{36}{17}$.

4 Solve the system $\begin{cases} 3x - 2y = 4 \\ 6x - 4y = 8 \end{cases}$. **The answer is $x = k$, $y = \dfrac{3k-4}{2}$.**

You've never seen this type of answer before, so stick with us and we'll explain what happened. Notice first of all that all you have to do is multiply the top equation by -2 to get $-6x + 4y = -8$, which is the exact opposite of the bottom equation, $6x - 4y = 8$. If you add these two you get $0 = 0$, which is always true. Therefore, this system has infinite solutions. Lots of answers will work in this system (actually, an infinite number of them). If you graph this system on a coordinate plane, you get two lines that lie on top of each other. How many points do those two lines share in common? All of them. Some books ask you to write this out using variables to represent constants. For example, if you arbitrarily pick that $x = k$, you can plug that into the top equation to get $3k - 2y = 4$, which means that $-2y = 4 - 3k$, or $y = \dfrac{3k-4}{2}$.

5 Solve $\begin{cases} x^2 - y = 1 \\ x + y = 5 \end{cases}$. **The answer is $x = 2$ and $y = 3$ or $x = -3$ and $y = 8$.**

As we explain in the section this problem was presented in, sometimes you end up with quadratics that, when solved, have two solutions. First, solve the linear equation for a variable, like x in the second equation: $x = 5 - y$. Now substitute this into the first equation: $(5 - y)^2 - y = 1$. FOIL out the binomial to get $25 - 10y + y^2 - y = 1$. Combine like terms: $25 - 11y + y^2 = 1$. Now get 0 on one side of the equation: $24 - 11y + y^2 = 0$. This factors to $(3 - y)(8 - y) = 0$, which, when you use the zero product property, gets you two solutions for y: $y = 3$ and $y = 8$. Accept that both of these are true and substitute them, one at a time, into the original quadratic equation to get the most *possible* solutions for x. First: If $y = 3$, then $x^2 - 3 = 1$; $x^2 = 4$; $x = \pm 2$. $2 + 3 = 5$ works in the second equation, but notice that $-2 + 3 = 5$ doesn't. That means that when y is 3, x only equals 3. Now do the same thing for $y = 8$: $x^2 - 8 = 1$; $x^2 = 9$; $x = \pm 3$. In the second equation, $3 + 8 = 5$ is false, but $-3 + 8 = 5$ is true, so the other solution is $x = -3$, $y = 8$.

6 Solve the system of equations $\begin{cases} x + y = 9 \\ xy = 20 \end{cases}$. **The answer is $x = 4$ and $y = 5$ or $x = 5$ and $y = 4$.**

First, solve the linear equation for x: $x = 9 - y$. Plug this into the second given equation: $(9 - y)y = 20$. Distribute to get $9y - y^2 = 20$ and get a quadratic to solve. Get 0 on one side: $0 = y^2 - 9y + 20$, which factors to $0 = (y - 5)(y - 4)$. This means that y is 5 or 4. Plug them both into either original equation: $y = 5$: $x + 5 = 9$, $x = 4$. $y = 4$: $x + 4 = 9$, $x = 5$.

7 Solve $\begin{cases} x^2 + y^2 = 1 \\ x + y^2 = -5 \end{cases}$. **The answer is no solution.**

This is also a first for you, but don't be surprised if your textbook or teacher throws these monkey wrenches at you, too. First, notice that both given equations have y^2 in them, with the same signs. If you multiply the second equation by -1, you get $\begin{cases} x^2 + y^2 = 1 \\ -x - y^2 = 5 \end{cases}$. Now add both equations together to get $x^2 - x = 6$. Next, get the equation to equal 0: $x^2 - x - 6 = 0$. This factors to $(x - 3)(x + 2) = 0$, which does give two solutions, $x = 3$ or $x = -2$. However, if you plug either one of these values into either of the original equations, you discover that y has no solution. This means that the system has no solution, either.

8 Solve $\begin{cases} 27x^2 - 16y^2 = -400 \\ -9x^2 + 4y^2 = 36 \end{cases}$. **The answer is $\left(\pm\dfrac{16}{3}, \pm\dfrac{25}{3} \right)$.**

The y terms have opposite signs, so it's easier to eliminate them after you multiply the second equation by 4: $\begin{cases} 27x^2 - 16y^2 = -400 \\ -36x^2 + 16y^2 = 144 \end{cases}$. Adding these two equations gets you that $-9x^2 = -256$, or $x^2 = {}^{256}\!/_9$, which finally means that $x = \pm {}^{16}\!/_3$. Now notice that both of the original equations have x^2 in them, but no x term. If you square ${}^{16}\!/_3$ or $-{}^{16}\!/_3$, you get the same result: ${}^{256}\!/_9$. This means the posi-tive and negative signs don't really matter when it comes to solving for y: $27\left(\dfrac{256}{9}\right) - 16y^2 = -400$.

Next, simplify: $768 - 16y^2 = -400$. Subtract 768 from both sides: $-16y^2 = -1168$. Now, by dividing -16, you get $y^2 = 73$, or $y = \pm\sqrt{73}$. Wow, that made even our heads hurt!

9 Solve $\begin{cases} \dfrac{14}{x+3} + \dfrac{7}{4-y} = 9 \\ \dfrac{21}{x+3} - \dfrac{3}{4-y} = 0 \end{cases}$. **The solution is $x = 4$ and $y = 3$.** First, rewrite the system by letting

$u = \dfrac{1}{x+3}$ and $v = \dfrac{1}{4-y}$ and getting $\begin{array}{l} 14u + 7v = 9 \\ 21u - 3v = 0 \end{array}$. Now, multiply the first equation by 3 and the

second equation by 7: $\begin{array}{l} 42u + 21v = 27 \\ 147u - 21v = 0 \end{array}$. Adding these two equations gets you $189u = 27$, which

means that $u = \frac{1}{7}$. Here's where you have to pay attention though! If this were a multiple choice test, most people would pick $\frac{1}{7}$ for the solution to this system, and they'd be wrong. That's why you have us, to help remind you that you're not done yet! That's because you found u but not x or y. Work your way backwards: $\dfrac{1}{7} = \dfrac{1}{x+3}$. If the numerators are equal, the denominators have

to be as well: $7 = x + 3$ means that $4 = x$. And you would have gotten the multiple choice correct! Now you can use that to get that $y = 3$ from the following steps:

$42\left(\dfrac{1}{7}\right) + 21v = 27$

$6 + 21v = 27$

$21v = 21$

$v = 1$

$1 = \dfrac{1}{4-y}$

$1 = 4 - y$

$y = 3$

Because this is a rational expression, also be sure to always check your solution to see if it's extraneous. In other words, if $x = 4$ or $y = 3$, do you get 0 in the denominator of either given equation? In this case, the answer is no — so these answers are legit!

10 Solve $\begin{cases} \dfrac{12}{x+1} - \dfrac{12}{y-1} = 8 \\ \dfrac{6}{x+1} - \dfrac{6}{y-1} = -2 \end{cases}$. **The answer is $x = 5$ and $y = -1$.**

If you let $u = \dfrac{1}{x+1}$ and $v = \dfrac{1}{y-1}$, you can rewrite the system as $\begin{array}{l} 12u - 12v = 8 \\ 6u + 6v = -2 \end{array}$. Now, multiply the

second equation by 2 and get $\begin{array}{l} 12u - 12v = 8 \\ 12u + 12v = -4 \end{array}$. Add them to get $24u = 4$, or $u = \frac{1}{6}$. Work your way

backwards from there: $\dfrac{1}{6} = \dfrac{1}{x+1}$; $6 = x + 1$; $x = 5$. The following steps get you to the solution for y:

$12\left(\dfrac{1}{6}\right) - 12v = 8$.

$2 - 12v = 8$

$-12v = 6$

$v = \dfrac{-1}{2}$

$\dfrac{-1}{2} = \dfrac{1}{y-1}$

$-2 = y - 1$

$-1 = y$

11 Solve the system $\begin{cases} 3x - 2y = 17 \\ x - 2z = 1 \\ 3y + 2z = 1 \end{cases}$. **The answer is $(x, y, z) = (5, -1, 2)$.**

A is the top equation: $3x - 2y = 17$. B is the middle equation: $x - 2z = 1$. C is $3y + 2z = 1$. C has no

x variable so you use the other two to eliminate x. A $- 3 \times$ B will do just that. $\begin{array}{c} 3x - 2y = 17 \\ -3(x - 2z = 1) \end{array}$

becomes $\begin{array}{c} 3x - 2y = 17 \\ -3x + 6z = -3 \end{array}$. Add these two equations to get $-2y + 6z = 14$, which you can call

equation D. 2C + 3D eliminates y: $\begin{array}{c} 2(3y + 2z = 1) \\ 3(-2y + 6z = 14) \end{array}$ becomes $\begin{array}{c} 6y + 4z = 2 \\ -6y + 18z = 42 \end{array}$. Add these two to

get $22z = 44$, or $z = 2$. Plug this into equation C: $3y + 2(2) = 1$; $3y + 4 = 1$; $3y = -3$; $y = -1$. Lastly, plug $z = 2$ into equation B: $x - 2(2) = 1$; $x - 4 = 1$; $x = 5$.

12 Solve $\begin{cases} 2x - y + z = 1 \\ x + y - z = 2 \\ -x - y + z = 2 \end{cases}$. **The answer is no solution.**

Right away, notice that all the coefficients on the middle and bottom equations are exact opposites of each other. When you add these two equations you get $0 = 4$. Because this equation is false, there's no solution.

13 Solve $\begin{cases} 2x + 3y + 4z = 37 \\ 4x - 3y + 2z = 17 \\ x + 2y - 3z = -5 \\ 3x - 2y + z = 11 \end{cases}$. **The answer is $x = 4$, $y = 3$, and $z = 5$.**

This system has three variables, so you only need three unique equations. Because you have four, you just have an additional one to play with. A: $2x + 3y + 4z = 37$. B: $4x - 3y + 2z = 17$. C: $x + 2y - 3z = -5$. D: $3x - 2y + z = 11$.

A + B eliminates y: $6x + 6z = 54$. Divide by 6 to get equation E: $x + z = 9$.

C + D also eliminates y: $4x - 2z = 6$. Divide by 2 to get equation F: $2x - z = 3$.

E + F eliminates z: $3x = 12$; $x = 4$.

Work backwards by plugging $x = 4$ into equation F: $2(4) - z = 3$; $8 - z = 3$; $-z = -5$; $z = 5$.

Now $x = 4$ and $z = 5$ go into equation A: $2(4) + 3y + 4(5) = 37$; $8 + 3y + 20 = 37$; $3y + 28 = 37$; $3y = 9$; $y = 3$.

14 Solve $\begin{cases} 3a + b + c + d = 0 \\ 4a + 5b + 2c = 15 \\ 4a + 2b + 5d = -10 \\ -5a + 3b - d = 8 \end{cases}$. **The answer is $a = 1$, $b = 3$, $c = -2$, and $d = -4$.**

To avoid confusion between A and a, we're naming the equations in this system a little differently:

T: $3a + b + c + d = 0$; S: $4a + 5b + 2c = 15$; N: $4a + 2b + 5d = -10$; L: $-5a + 3b - d = 8$. Now start eliminating!

2T + S: $\begin{cases} -2(3a + b + c + d = 0) \\ 4a + 5b + 2c = 15 \end{cases} \rightarrow \begin{cases} -6a - 2b - 2c - 2d = 0 \\ 4a + 5b + 2c = 15 \end{cases}$ gives you equation M: $-2a + 3b - 2d =$

15. That way, you now have three equations with a, b, and d in them: N, L, and the new one, M. Start over and eliminate another variable.

$5M + 2N$: $\begin{Bmatrix} 5(-2a + 3b - 2d = 15) \\ 2(4a + 2b + 5d = -10) \end{Bmatrix} \rightarrow \begin{Bmatrix} -10a + 15b - 10d = 75 \\ 8a + 4b + 10d = -20 \end{Bmatrix}$ gives you equation R: $-2a + 19b = 55$.

$M - 2L$: $\begin{Bmatrix} -2a + 3b - 2d = 15 \\ -2(-5a + 3b - d = 8) \end{Bmatrix} \rightarrow \begin{Bmatrix} -2a + 3b - 2d = 15 \\ 10a - 6b + 2d = -16 \end{Bmatrix}$ gives you equation P: $8a - 3b = -1$. You're

down to two equations and two variables (happy dance)!

$4R + P$: $\begin{Bmatrix} 4(-2a + 19b = 55) \\ 8a - 3b = -1 \end{Bmatrix} \rightarrow \begin{Bmatrix} -8a + 76b = 220 \\ 8a - 3b = -1 \end{Bmatrix}$. Add these two equations to get $73b = 219$. This

gives you your first solution, $b = 3$. Work your way backwards to get the other solutions.

P: $8a - 3(3) = -1$; $8a - 9 = -1$; $8a = 8$; $a = 1$.

M: $-2(1) + 3(3) - 2d = 15$; $-2 + 9 - 2d = 15$; $7 - 2d = 15$; $-2d = 8$; $d = -4$.

T: $3(1) + 3 + c - 4 = 0$; $3 + 3 + c - 4 = 0$; $2 + c = 0$; $c = -2$.

15 Sketch the graph of $\begin{cases} 2x + y \ge 9 \\ 2x - y \ge 1 \\ x \le 7 \end{cases}$ **See the graph for the answer.**

Put the first two equations in slope-intercept form first. The top equation is $y \ge -2x + 9$; the second equation is $y \le 2x - 1$. Put them all on the same graph.

16 Sketch the graph of $\begin{cases} x^2 + y^2 \ge 9 \\ x^2 + (y - 3)^2 \ge 9 \end{cases}$. **See the graph for the answer.**

Both of these equations are circles. If you don't recognize them as such, you should turn to Chapter 12 to read up on conic sections. Graph them both on the same graph.

17 Sketch the graph of $\begin{cases} x^2 - y > 2 \\ x - y < 4 \end{cases}$. **See the graph for the answer.**

The first equation is a parabola, $x^2 - 2 > y$. The second equation is the line $x - 4 < y$. Graph them both on the same graph.

18 Sketch the graph of $\begin{cases} x \ge 0 \\ x + y < 4 \\ y \le \sqrt{x-1} \end{cases}$. **See the graph for the answer.**

The third equation is a square root function. If you don't remember how to graph it, turn to Chapter 3 and get a refresher.

19 Find the constants A and B: $\dfrac{x - 38}{x^2 + x - 12} = \dfrac{A}{x + 4} + \dfrac{B}{x - 3}$. **The answer is A = 6 and B = –5.**

We've just started the decomposition process for you. The first thing you should do is multiply everything by the factored denominator

$$\frac{x - 38}{(x+4)(x-3)}(x+4)(x-3) = \frac{A}{x+4}(x+4)(x-3) + \frac{B}{x-3}(x+4)(x-3); \text{ in other words,}$$

$x - 38 = A(x - 3) + B(x + 4)$. Multiply everything out to get $x - 38 = Ax - 3A + Bx + 4B$. Collect the like terms on the right: $x - 38 = Ax + Bx - 3A + 4B$. Now factor out the x: $x - 38$ (A + B)x – 3A + 4B. The coefficients of the x terms are equal, which gives you one equation: 1 = A + B. The constants are also equal, which gives you a second equation: –38 = –3A + 4B. This system of equa-

tions is solved as follows: $\begin{cases} 3(1 = A + B) \\ -38 = -3A + 4B \end{cases} \rightarrow \begin{cases} 3 = 3A + 3B \\ -38 = -3A + 4B \end{cases}$. Adding these two equations

gives you –35 = 7B, or –5 = B. Now, 1 = A – 5 tells you that 6 = A.

20 Find the form of the partial fraction decomposition for $\dfrac{5x-4}{\left(x-1\right)^2}$, but don't find the constants.

The answer is $\dfrac{5x-4}{\left(x-1\right)^2} = \dfrac{A}{x-1} + \dfrac{B}{\left(x-1\right)^2}$.

Note the directions here. We just ask you to set up the problem, not to perform the decomposition. The denominator is already factored for you. The hardest part about this problem is to remember that you need to use one term for every power of this binomial, up to its degree of 2. In other words, you need one fraction with $(x-1)^1$ in it and you need another with $(x-1)^2$ in it. This is why the answer is $\dfrac{5x-4}{\left(x-1\right)^2} = \dfrac{A}{x-1} + \dfrac{B}{\left(x-1\right)^2}$.

21 Find the partial fraction decomposition for $\dfrac{2x^2 - 21x + 18}{\left(x-1\right)\left(x^2 - 4x + 4\right)}$. **The answer is**

$\dfrac{-1}{x-1} + \dfrac{3}{x-2} - \dfrac{16}{\left(x-2\right)^2}$.

First, factor the given denominator to become $\dfrac{2x^2 - 21x + 18}{\left(x-1\right)\left(x-2\right)^2}$. Because each factor is linear, set

up three different fractions with constants on the top: one for the $(x-1)$ factor, one for the $(x-2)$ factor, and one for the $(x-2)^2$ factor. $\dfrac{2x^2 - 21x + 18}{\left(x-1\right)\left(x-2\right)^2} = \dfrac{A}{x-1} + \dfrac{B}{x-2} + \dfrac{C}{\left(x-2\right)^2}$.

Multiply every term by the factored denominator, cancel, and get $2x^2 - 21x + 18 = A(x-2)^2 + B(x-1)(x-2) + C(x-1)$. Multiply this all out to get $2x^2 - 21x + 18 = Ax^2 - 4Ax + 4A + Bx^2 - 3Bx + 2B + Cx - C$. Collect the like terms: $2x^2 - 21x + 18 = Ax^2 + Bx^2 - 4Ax - 3Bx + Cx + 4A + 2B - C$. Factor out the x^2 and the x on the right side: $2x^2 - 21x + 18 = (A + B)x^2 + (-4A - 3B + C)x + 4A + 2B - C$. Set the coefficients of x^2 equal to each other: $2 = A + B$. Set the coefficients of x equal to each other: $-21 = -4A - 3B + C$. Lastly, set the constants equal to each other: $18 = 4A + 2B - C$. Solve this system of equations using elimination to get $A = -1$, $B = 3$, and $C = -16$. This is where the answer $\dfrac{-1}{x-1} + \dfrac{3}{x-2} - \dfrac{16}{\left(x-2\right)^2}$ comes from.

22 Find the partial fraction decomposition for $\dfrac{11x^2 - 7x + 14}{2x^3 - 4x^2 + 3x - 6}$. **The answer is** $\dfrac{3x-1}{2x^2+3} + \dfrac{4}{x-2}$.

Factor the denominator by using grouping: $2x^2(x-2) + 3(x-2)$ becomes $(x-2)(2x^2+3)$. Use this to make two different fractions with a constant on top of the linear factor and a linear expression on top of the quadratic factor: $\dfrac{11x^2 - 7x + 14}{2x^3 - 4x^2 + 3x - 6} = \dfrac{Ax+B}{2x^2+3} + \dfrac{C}{x-2}$. Multiply every

term by the factored denominator, cancel, and get $11x^2 - 7x + 14 = (Ax+B)(x-2) + C(2x^2+3)$. Multiply it all out to get $11x^2 - 7x + 14 = Ax^2 - 2Ax + Bx - 2B + 2Cx^2 + 3C$. Collect the like terms and factor to get $11x^2 - 7x + 14 = (A + 2C)x^2 + (-2A + B)x - 2B + 3C$. This gives you a system with

three equations: $\begin{cases} 11 = A + 2C \\ -7 = -2A + B \\ 14 = -2B + 3C \end{cases}$. Solving this system tells you that $A = 3$, $B = -1$, and $C = 4$,

which gives you the answer $\dfrac{3x-1}{2x^2+3} + \dfrac{4}{x-2}$.

23 Find 4N. **The answer is** $\begin{bmatrix} 8 & 16 & 20 & -32 \\ 40 & 12 & -8 & -12 \end{bmatrix}$.

First, write out the problem by substituting the given matrix N in $4\begin{bmatrix} 2 & 4 & 5 & -8 \\ 10 & 3 & -2 & -3 \end{bmatrix}$. Distribute

the 4 to every element inside the matrix to get the answer: $\begin{bmatrix} 8 & 16 & 20 & -32 \\ 40 & 12 & -8 & -12 \end{bmatrix}$.

24 Find 4N + 5M. **The answer is** $\begin{bmatrix} -17 & 11 & 35 & -2 \\ 40 & 22 & -18 & 18 \end{bmatrix}$.

First, substitute the given matrices into the expression: $4\begin{bmatrix} 2 & 4 & 5 & -8 \\ 10 & 3 & -2 & -3 \end{bmatrix} + 5\begin{bmatrix} -5 & -1 & 3 & 6 \\ 0 & 2 & -2 & 6 \end{bmatrix}$.

Distribute both scalars to every element of their matrix: $\begin{bmatrix} 8 & 16 & 20 & -32 \\ 40 & 12 & -8 & -12 \end{bmatrix} + \begin{bmatrix} -25 & -5 & 15 & 30 \\ 0 & 10 & -10 & 30 \end{bmatrix}$.

Add these two matrices by adding their corresponding elements to get $\begin{bmatrix} -17 & 11 & 35 & -2 \\ 40 & 22 & -18 & 18 \end{bmatrix}$.

25 Find 3M – P. **The answer is no solution.**

These matrices aren't the same dimensions, so you can't add them. There's no solution.

26 Find NP. **The answer is** $\begin{bmatrix} 64 & 19 & 11 \\ 13 & 20 & -9 \end{bmatrix}$.

Substitute the given matrices into the expression $\begin{bmatrix} 2 & 4 & 5 & -8 \\ 10 & 3 & -2 & -3 \end{bmatrix}\begin{bmatrix} -1 & 2 & -1 \\ 4 & 4 & 0 \\ 2 & 3 & 1 \\ -5 & 2 & -1 \end{bmatrix}$. Check to see if

you can even multiply them. The matrix on the left is 2×4 and the one on the right is 4×3, so you can multiply them. Multiply every row from the left matrix by every column from the right matrix.

The sum of the first row times the first column: $2(-1) + 4(4) + 5(2) - 8(-5) = 64$.

The sum of the first row times the second column: $2(2) + 4(4) + 5(3) - 8(2) = 19$.

The sum of the first row times the third column: $2(-1) + 4(0) + 5(1) - 8(-1) = 11$.

The sum of the second row times the first column: $10(-1) + 3(4) - 2(2) - 3(-5) = 13$.

The sum of the second row times the second column: $10(2) + 3(4) - 2(3) - 3(2) = 20$.

The sum of the second row times the third column: $10(-1) + 3(0) - 2(1) - 3(-1) = -9$.

Putting these all into a matrix gives you the answer: $\begin{bmatrix} 64 & 19 & 11 \\ 13 & 20 & -9 \end{bmatrix}$.

27 Using the augmented matrix from the last example, use elementary row operations to find

$-3r_2 \rightarrow r_2$. **The answer is** $\left[\begin{array}{cc|c} 3 & -1 & 6 \\ -6 & -9 & -9 \end{array}\right]$.

Just multiply the second row of the given equation by –3 to get $\left[\begin{array}{cc|c} 3 & -1 & 6 \\ -6 & -9 & -9 \end{array}\right]$.

28 Now, using your answer from problem 27, find $r_1 \leftrightarrow r_2$. **The answer is** $\begin{bmatrix} -6 & -9 & \vline & -9 \\ 3 & -1 & \vline & 6 \end{bmatrix}$.

Swap the first row with the second row and get $\begin{bmatrix} -6 & -9 & \vline & -9 \\ 3 & -1 & \vline & 6 \end{bmatrix}$.

29 Now, keep going and find $r_1 + r_2 \rightarrow r_2$. **The answer is** $\begin{bmatrix} -6 & -9 & \vline & -9 \\ -3 & -10 & \vline & -3 \end{bmatrix}$.

Add each element from row one to its corresponding element in row two to change row two and get $\begin{bmatrix} -6 & -9 & \vline & -9 \\ -3 & -10 & \vline & -3 \end{bmatrix}$.

30 Lastly, find $3r_2 + r_1 \rightarrow r_1$. **The answer is** $\begin{bmatrix} -15 & -39 & \vline & -18 \\ -3 & -10 & \vline & -3 \end{bmatrix}$.

Temporarily multiply the second row by 3 to get $[-9\ -30\ |\ -9]$. Add these to the corresponding elements in row one to change row one and get $\begin{bmatrix} -15 & -39 & \vline & -18 \\ -3 & -10 & \vline & -3 \end{bmatrix}$.

31 Solve the system of equations $\begin{cases} 2x + 5y = 7 \\ 3x - 5y = 2 \end{cases}$ by writing it in augmented form and then putting the matrix in row echelon form. **The answer is $x = \frac{9}{5}$ and $y = \frac{17}{25}$.**

The matrix in augmented form is $\begin{bmatrix} 2 & 5 & \vline & 7 \\ 3 & -5 & \vline & 2 \end{bmatrix}$. Multiply the top row by $\frac{1}{2}$ to get a 1 in the upper left corner: $\begin{bmatrix} 1 & \frac{5}{2} & \vline & \frac{7}{2} \\ 3 & -5 & \vline & 2 \end{bmatrix}$. You need to add a -3 to the second row to get a 0 under the one:

$-3r_1 + r_2 \rightarrow r_2$ gives you $\begin{bmatrix} 1 & \frac{5}{2} & \vline & \frac{7}{2} \\ 0 & \frac{-25}{2} & \vline & \frac{-17}{2} \end{bmatrix}$; this is in row echelon form. Set up an equation from the second row: $\frac{-25}{2}y = \frac{-17}{2}$. Solve this equation to get $y = \frac{17}{25}$. Use that answer and back substitute $x + \frac{5}{2}\left(\frac{17}{25}\right) = \frac{7}{2}$ to get that $x = \frac{9}{5}$.

32 Use Gaussian elimination to solve $\begin{cases} 3x - 2y + 6z = 7 \\ x - 2y - z = -2 \\ -3x + 10y + 11z = 18 \end{cases}$. **The answer is $x = 1$, $y = 1$, and $z = 1$.**

Set up the system as an augmented matrix: $\begin{bmatrix} 3 & -2 & 6 & \vline & 7 \\ 1 & -2 & -1 & \vline & -2 \\ -3 & 10 & 11 & \vline & 18 \end{bmatrix}$.

$r_1 \leftrightarrow r_2$ gets a 1 in the upper-left corner: $\begin{bmatrix} 1 & -2 & -1 & \vline & -2 \\ 3 & -2 & 6 & \vline & 7 \\ -3 & 10 & 11 & \vline & 18 \end{bmatrix}$.

$-3r_1 + r_2 \rightarrow r_2$ gets a 0 under the 1 in the second row: $\begin{bmatrix} 1 & -2 & -1 & \vline & -2 \\ 0 & 4 & 9 & \vline & 13 \\ -3 & 10 & 11 & \vline & 18 \end{bmatrix}$.

$3r_1 + r_3 \rightarrow r_3$ gets a 0 under the 1 in the third row: $\begin{bmatrix} 1 & -2 & -1 & | & -2 \\ 0 & 4 & 9 & | & 13 \\ 0 & 4 & 8 & | & 12 \end{bmatrix}$.

$\frac{1}{4}r_2 \rightarrow r_2$ gets a 1 in the next position along the diagonal: $\begin{bmatrix} 1 & -2 & -1 & | & -2 \\ 0 & 1 & 9/4 & | & 13/4 \\ 0 & 4 & 8 & | & 12 \end{bmatrix}$.

$-4r_2 + r_3 \rightarrow r_3$ gets a 0 under the 1 you just created: $\begin{bmatrix} 1 & -2 & -1 & | & -2 \\ 0 & 1 & 9/4 & | & 13/4 \\ 0 & 0 & -1 & | & -1 \end{bmatrix}$. This means that $-z = -1$,

or $z = 1$. Back substitute $y + \frac{9}{4}(1) = \frac{13}{4}$ to get that $y = 1$. Back substitute again: $x - 2(1) - 1(1) = -2$, or $x = 1$.

33 Solve the system $\begin{cases} 4x - y = -10 \\ 2x + 3y = 16 \end{cases}$ using inverse matrices. **The answer is $x = -1$ and $y = 6$.**

First, write the system as a matrix equation: $\begin{bmatrix} 4 & -1 \\ 2 & 3 \end{bmatrix}\begin{bmatrix} x \\ y \end{bmatrix} = \begin{bmatrix} -10 \\ 16 \end{bmatrix}$. Now, find the inverse matrix

using the handy formula we show you in the section in which this question appears:

$\frac{1}{4(3) - (-1)(2)}\begin{bmatrix} 3 & 1 \\ -2 & 4 \end{bmatrix} = \frac{1}{14}\begin{bmatrix} 3 & 1 \\ -2 & 4 \end{bmatrix}$. Multiply this inverse on both sides of the equation:

$\frac{1}{14}\begin{bmatrix} 3 & 1 \\ -2 & 4 \end{bmatrix}\begin{bmatrix} 4 & -1 \\ 2 & 3 \end{bmatrix}\begin{bmatrix} x \\ y \end{bmatrix} = \frac{1}{14}\begin{bmatrix} 3 & 1 \\ -2 & 4 \end{bmatrix}\begin{bmatrix} -10 \\ 16 \end{bmatrix}$. Multiply the two matrices: $\begin{bmatrix} x \\ y \end{bmatrix} = \frac{1}{14}\begin{bmatrix} -14 \\ 84 \end{bmatrix}$. Multiply

the scalar: $\begin{bmatrix} x \\ y \end{bmatrix} = \begin{bmatrix} -1 \\ 6 \end{bmatrix}$. Your solutions from top to bottom are $x = -1$ and $y = 6$.

34 Solve the system $\begin{cases} 4x + 3y = 17 \\ 2x - y = 11 \end{cases}$ using inverse matrices. **The answer is $x = 5$ and $y = -1$.**

Write the system as a matrix equation: $\begin{bmatrix} 4 & 3 \\ 2 & -1 \end{bmatrix}\begin{bmatrix} x \\ y \end{bmatrix} = \begin{bmatrix} 17 \\ 11 \end{bmatrix}$. Find the inverse:

$\frac{1}{4(-1) - 2(3)}\begin{bmatrix} -1 & -3 \\ -2 & 4 \end{bmatrix} = \frac{-1}{10}\begin{bmatrix} -1 & -3 \\ -2 & 4 \end{bmatrix}$. Multiply this on both sides:

$\frac{-1}{10}\begin{bmatrix} -1 & -3 \\ -2 & 4 \end{bmatrix}\begin{bmatrix} 4 & 3 \\ 2 & -1 \end{bmatrix}\begin{bmatrix} x \\ y \end{bmatrix} = \frac{-1}{10}\begin{bmatrix} -1 & -2 \\ -2 & 4 \end{bmatrix}\begin{bmatrix} 17 \\ 11 \end{bmatrix}$. Multiply the matrices: $\begin{bmatrix} x \\ y \end{bmatrix} = \frac{-1}{10}\begin{bmatrix} -50 \\ 10 \end{bmatrix}$. Multiply the

scalar: $\begin{bmatrix} x \\ y \end{bmatrix} = \begin{bmatrix} 5 \\ -1 \end{bmatrix}$.

35 Find the determinant of $\begin{vmatrix} 2 & -1 & 4 \\ -3 & 4 & 6 \\ -2 & -1 & 5 \end{vmatrix}$. **The answer is 93.**

Because this is a 3×3 matrix, you have to use diagonals. First, rewrite the first two columns

after the third one: $\begin{vmatrix} 2 & -1 & 4 \\ -3 & 4 & 6 \\ -2 & -1 & 5 \end{vmatrix}\begin{matrix} 2 & -1 \\ -3 & 4 \\ -2 & -1 \end{matrix}$. The sum of the diagonals from bottom-left to top-right is: $(-2)(4)(4) + (-1)(6)(2) + (5)(-3)(-1) = -32 - 12 + 15 = -29$. The sum of the diagonals from top-left to bottom-right is: $(2)(4)(5) + (-1)(6)(-2) + 4(-3)(-1) = 40 + 12 + 12 = 64$. The difference of the bottom sum minus the top sum is $64 - (-29) = 93$.

36 Use Cramer's Rule to solve $\begin{cases} 2x - y + 4z = 7 \\ -3x + 4y + 6z = -1 \\ -2x - y + 5z = 4 \end{cases}$ **The answer is $x = 1$, $y = -1$, and $z = 1$.**

Set up the quotient to find x first: $x = \dfrac{\begin{vmatrix} 7 & -1 & 4 \\ -1 & 4 & 6 \\ 4 & -1 & 5 \end{vmatrix}}{\begin{vmatrix} 2 & -1 & 4 \\ -3 & 4 & 6 \\ -2 & -1 & 5 \end{vmatrix}} = \dfrac{93}{93} = 1$. We use an online matrix calculator to

find the determinants here. Out of all the tips we could give you, this is probably the best one. Learn how to use your calculator or the Internet to calculate matrix operations like a determinant. If your teacher doesn't allow you to use a calculator on a test, you have to use the process of diagonals to find the determinant here.

Solve for y: $y = \dfrac{\begin{vmatrix} 2 & 7 & 4 \\ -3 & -1 & 6 \\ -2 & -4 & 5 \end{vmatrix}}{\begin{vmatrix} 2 & -1 & 4 \\ -3 & 4 & 6 \\ -2 & -1 & 5 \end{vmatrix}} = \dfrac{-93}{93} = -1$. Ditto.

Solve for z: $z = \dfrac{\begin{vmatrix} 2 & -1 & 7 \\ -3 & 4 & -1 \\ -2 & -1 & 4 \end{vmatrix}}{\begin{vmatrix} 2 & -1 & 4 \\ -3 & 4 & 6 \\ -2 & -1 & 5 \end{vmatrix}} = \dfrac{93}{93} = 1$. Ditto squared.

Chapter 14

Sequences, Series, and Binomials — Oh My!

This chapter is all about patterns. No, we're not making quilts, although we could . . . nah! We'll stick with patterns of numbers, not cloth. Namely, we explore sequences, series, and the binomial theorem.

A *sequence* is an ordered list of numbers that follow a pattern. A *series* is the sum of the terms in a sequence, so it adds a pattern. The *binomial theorem* is the result of discovering the pattern of an expanded binomial.

Hmm . . . we think we're sensing a pattern of patterns here.

One mathematical term that comes up in this chapter is *factorial,* which you may remember from your previous math classes. A factorial, n!, read "n factorial," is defined as $1 \cdot 2 \cdot 3 \cdot \ldots \cdot (n-1) \cdot n$.

Major General Sequences and Series: Calculating Terms

Mathematically, a sequence is usually written in the following form: $\{a_n\} = a_1, a_2, a_3, \ldots a_n$. Here, *n* is the number of terms, a_n is the *term of the sequence,* and a_1 is the first term, a_2 is the second term, and so on. Similarly, a series can be written as the sum of the terms: $a_1 + a_2 + a_3 + \ldots + a_n$. The pattern of sequences and series can usually be described by a general expression or rule. Because sequences and series can be infinite, this expression allows you to find any number in the list without having to find all the numbers. If you're not given the general expression, you can find it if you're given the first few terms of a sequence or series.

Sometimes a term in a sequence depends on the term(s) before it. These are called *recursive sequences.* A famous example of a recursive sequence is the Fibonacci Sequence: 1, 1, 2, 3, 5, 8, 13, . . . where each term is the sum of the two before it.

Q. Write the first five terms of the sequence $a_n = n^2 + 3$.

A. **4, 7, 12, 19, 28.** To find each term, you just plug the number of the term (n) into the formula: $a_1 = (1)^2 + 3 = 4$; $a_2 = (2)^2 + 3 = 7$; $a_3 = (3)^2 + 3 = 12$; $a_4 = (4)^2 + 3 = 19$; $a_5 = (5)^2 + 3 = 28$.

Q. Write a general expression for the sequence to find the nth term: $-\frac{1}{2}, \frac{2}{3}, -\frac{3}{4}, \frac{4}{5}$.

A. $a_n = (-1)^n \dfrac{n}{n+1}$. First, notice that the sign alternates between negative and positive. To deal with this, multiply by powers of -1: $(-1)^n$. Next, notice that the sequence's numerator is the same as the term number (n), so n becomes your numerator. Finally, you can see that the denominator is simply one number larger than the numerator (and term number), so it can be written as $n + 1$. Putting these pieces together, you get: $a_n = (-1)^n \dfrac{n}{n+1}$.

---

1. Write the first five terms of the sequence:
$$a_n = \frac{n-3}{2n}.$$

Solve It

2. Write a general expression for the sequence to find the nth term: 2, 4, 10, 28, 82, . . .

Solve It

---

3. Write a general expression for the sequence to find the nth term: 2, 2, $\frac{8}{3}$, 4, $\frac{32}{5}$, . . .

Solve It

4. Write the next two terms of the sequence: 1, 1, 2, 3, 7, 16.

Solve It

5. Find the sum of the first five terms of the series: $2^{n-1} + 1$.

Solve It

6. Find the sum of the first five terms of the series: $3^n + 2n$.

Solve It

Working Out the Common Difference: Arithmetic Sequences and Series

One special type of sequence is called an *arithmetic sequence*. In these sequences, each term differs from the one before it by a *common difference*, d. As a result, you have a formula for finding the nth term of an arithmetic sequence:

$$a_n = a_1 + (n-1)d$$

where a_1 is the first term, n is the number of terms, and d is the common difference.

To find the sum of an arithmetic sequence, also called an *arithmetic series,* you have to add a given number of terms together. This you can write in summation notation:

$$S_k = \sum_{n=1}^{k} a_n = \frac{k}{2}\left(a_1 + a_k\right)$$

This is read as "the kth partial sum of a_n" where $n = 1$ is the lower limit, k is the sum's upper limit, a_1 is the first term, and a_k is the last term. To find a sum, simply plug the lower and upper limits into the formula for a_n to find a_1 and a_k and then simplify.

Q. Find the 60th term of the arithmetic sequence: 4, 7, 10, 13, . . .

A. $a_{60} = 181$. The easiest way to begin this problem is to find the formula for the nth term. To do so, you need a_1, which is 4. You also need the common difference, d, which can be found by subtracting two sequential terms; for example: $a_2 - a_1 = 7 - 4 = 3$. Plugging these into the general formula and simplifying, you get: $a_n = a_1 + (n-1)d = 4 + (n-1)3 = 4 + 3n - 3 = 1 + 3n$. Now you can find the 60th term by plugging in 60 for n: $a_{60} = 1 + 3(60) = 1 + 180 = 181$.

Q. Find $\sum_{n=1}^{8} -3n + 5$.

A. **−68.** To find the sum, you just have to use the arithmetic series formula. For this, you need k (which is 8), a_1, and a_k. Start by finding a_1: $-3(1) + 5 = 2$. Then find a_8: $-3(8) + 5 = -19$. Finally, plug these into the formula: $S_n = (\frac{8}{2})(2 + -19) = (4)(-17) = -68$.

7. Find the 50th term of the arithmetic sequence: $-6, -1, 4, 9, \ldots$

Solve It

8. Find the general formula of an arithmetic sequence where $a_1 = -3$ and $a_{15} = 53$.

Solve It

9. Find the general formula of an arithmetic sequence where $a_5 = -5$ and $a_{20} = -35$.

Solve It

10. Find $\displaystyle\sum_{n=1}^{5} \frac{1}{2}n + 2$.

Solve It

11. Find $\displaystyle\sum_{n=4}^{10} 2n - 3$.

Solve It

12. Write the arithmetic series $2 + \frac{7}{3} + \frac{8}{3} + 3 + \frac{10}{3}$ in summation notation and find the result.

Solve It

Be Fruitful and Multiply: Simplifying Geometric Sequences and Series

When consecutive terms in a sequence have a common ratio, the sequence is called a *geometric sequence*. To find that ratio, r, you divide each term by the term before it, and the quotient should be the same. Just like the other sequences, a_1 denotes the first term. To find the next term, multiply by the common ratio, r. Another pattern! The formula for the nth term of a geometric sequence is:

$$a_n = a_1 \cdot r^{n-1}$$

As with other sequences, you can find the sum of geometric sequences, called *geometric series*. To find a partial sum of a geometric sequence you can use the following formula:

$$S_n = \sum_{n=1}^{k} a_n = a_1 \left(\frac{1-r^k}{1-r} \right)$$

Here, $n = 1$ is the lower limit, k is the sum's upper limit, r is the common ratio, and a_1 is the first term. To find a sum, simply plug the lower and upper limits into the formula for a_n to find a_1 and a_k and then simplify.

Unlike any other sequence, you can actually find the value of an infinite sum of many geometric sequences. As long as r lies within the range $-1 < r < 1$, you can find the infinite sum. If r lies outside that range, a_n will grow infinitely, so the sum won't have a limit. To find the infinite sum of a geometric series where r is within the range $-1 < r < 1$, use the following formula:

$$S = \sum_{n=1}^{\infty} a_n = \frac{a_1}{1-r}$$

Because we just plug and chug, geometric sequences and series are pretty easy to deal with. You just need to remember your rules for simplifying fractions. You can do it!

Q. Find the 10th term of the geometric sequence: 3, –6, 12, –24, . . .

A. $a_{10} = -1{,}536$. For the formula for the nth term of a geometric sequence, you need a_1 and r. a_1 is given in the problem: 3. To find r, all you need to do is divide a_2 by a_1: $-6/3 = -2$. Now you can simply plug these values into the formula: $a_{10} = 3 \cdot (-2)^{10-1} = 3 \cdot (-2)^9 = 3(-512) = -1{,}536$.

Q. Find the sum: $\sum_{n=1}^{5} 6 \left(\frac{1}{3} \right)^{n-1}$.

A. $242/27$. To use the partial sum formula, you need to know a_1, r, and k. From the problem, you can identify r as $\frac{1}{3}$ and k as 5. To find a_1, simply plug in 1 for n: $a_1 = 6(\frac{1}{3})^{1-1} = 6 \cdot (\frac{1}{3})^0 = 6 \cdot 1 = 6$. Now all you have to do is plug and chug:

$$S_5 = 6 \left(\frac{1 - \left(\frac{1}{3} \right)^5}{1 - \frac{1}{3}} \right) = 6 \cdot \frac{\frac{242}{243}}{\frac{2}{3}} = 6 \cdot \frac{242}{243} \cdot \frac{3}{2} = 242/27.$$

13. Find the 16th term of a geometric sequence given $a_1 = 5$ and $a_2 = -15$.

Solve It

14. Find the 8th term of a geometric sequence given $a_2 = 6$ and $a_6 = 486$.

Solve It

15. Find the sum $\sum_{n=1}^{\infty} 3\left(-\frac{2}{3}\right)^{n-1}$.

Solve It

16. Find the partial sum of the geometric series: $\frac{1}{6} + \frac{1}{3} + \frac{2}{3} + \ldots \frac{32}{3}$.

Solve It

17. Find the sum of the infinite geometric series: $\frac{1}{6} + \frac{1}{3} + \frac{2}{3} + \ldots$.

Solve It

18. Find the sum: $\sum_{n=1}^{\infty} 3\left(-\frac{2}{3}\right)^{n-1}$.

Solve It

Expanding Polynomials Using the Binomial Theorem

Binomials are polynomials with exactly two terms. Often, binomials are raised to powers to complete computations, and when you multiply out a binomial so that it doesn't have any parentheses, it's called a *binomial expansion*. One way to complete binomial expansions is to distribute terms, but if the power is high, this method can be tedious.

An easier way to expand binomials is to use the *binomial theorem*:

$$(a+b)^n = \binom{n}{0}a^n b^0 + \binom{n}{1}a^{n-1}b^1 + \binom{n}{2}a^{n-2}b^2 + \ldots + \binom{n}{n-2}a^2 b^{n-2} + \binom{n}{n-1}a^1 b^{n-1} + \binom{n}{n}a^0 b^n$$

Here, a is the first term, b is the second term, and $\binom{n}{r} = \dfrac{n!}{r!(n-r)!}$ is the *combinations*

formula. For example, to find the binomial coefficient given by $\binom{6}{2}$, plug the values into

the formula and simplify: $\dfrac{6!}{2!(6-2)!} = \dfrac{6!}{2!4!} = \dfrac{720}{2 \cdot 24} = 15$.

This seems like a lot of work, but trust us — if you just take it one step at a time, this method will save you an immense amount of time!

Q. Write the expansion of $(3x-2)^4$.

A. $81x^4 - 216x^3 + 216x^2 - 96x + 16$. To expand, simply replace a with $3x$, b with -2, and n with 4 to get

$\binom{4}{0}(3x)^4(-2)^0 + \binom{4}{1}(3x)^3(-2)^1 + \binom{4}{2}(3x)^2(-2)^2 + \binom{4}{3}(3x)^1(-2)^3 + \binom{4}{4}(3x)^0(-2)^4$. Now, to simplify

this mess, start with the combinations formula for each term:

$(1)(3x)^4(-2)^0 + (4)(3x)^3(-2)^1 + (6)(3x)^2(-2)^2 + (4)(3x)^1(-2)^3 + (1)(3x)^0(-2)^4$.

Then, raise the monomials to the specified powers:

$(1)(81x^4)(1) + (4)(27x^3)(-2) + (6)(9x^2)(4) + (4)(3x)(-8) + (1)(1)(16)$.

Finally, combine like terms and simplify: $81x^4 - 216x^3 + 216x^2 - 96x + 16$.

19. Find the 5th term of $(x + y)^{12}$.

Solve It

20. Find the 8th term of $(2x - 3y)^{10}$.

Solve It

21. Expand $(k - 4)^5$.

Solve It

22. Expand $(y + 4z)^6$.

Solve It

Answers to Problems on Sequences, Series, and Binomials

1 Write the first five terms of the sequence: $a_n = \dfrac{n-3}{2n}$.

The answer is –1, –¼, 0, ⅛, ⅕. To find each term, you plug the number of the term (n) into the formula:
$$a_1 = \frac{1-3}{2 \cdot 1} = \frac{-2}{2} = -1, \quad a_2 = \frac{2-3}{2 \cdot 2} = \frac{-1}{4}, \quad a_3 = \frac{3-3}{2 \cdot 3} = \frac{0}{6} = 0, \quad a_4 = \frac{4-3}{2 \cdot 4} = \frac{1}{8}, \quad a_5 = \frac{5-3}{2 \cdot 5} = \frac{2}{10} = \frac{1}{5}$$
$= -1, -¼, 0, ⅛, ⅕.$

2 Write a general expression for the sequence to find the nth term: 2, 4, 10, 28, 82, . . .

The answer is $a_n = 3^{n-1} + 1$. At first, this one seems tricky. But if you subtract 1 from each term, the pattern becomes apparent: 1, 3, 9, 27, 81 . . . these are multiples of 3! In fact, they can be found by taking 3 to sequential powers: $3^0, 3^1, 3^2, 3^3, 3^4$, where the power is 1 less than the term, or $n - 1$. So, putting that all together, we get $a_n = 3^{n-1} + 1$.

3 Write a general expression for the sequence to find the nth term: 2, 2, ⅔, 4, ³²⁄₅, . . .

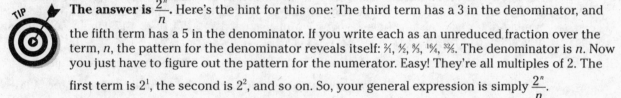

The answer is $\dfrac{2^n}{n}$. Here's the hint for this one: The third term has a 3 in the denominator, and the fifth term has a 5 in the denominator. If you write each as an unreduced fraction over the term, n, the pattern for the denominator reveals itself: ²⁄₁, ⁴⁄₂, ⁸⁄₃, ¹⁶⁄₄, ³²⁄₅. The denominator is n. Now you just have to figure out the pattern for the numerator. Easy! They're all multiples of 2. The first term is 2^1, the second is 2^2, and so on. So, your general expression is simply $\dfrac{2^n}{n}$.

4 Write the next two terms of the sequence: 1, 1, 2, 3, 7, 16, . . .

The answer is 65, 321. Ah! This is one of those recursive sequences we were mentioning. This pattern is found by adding the first term squared to the second term, and so on: $1^2 + 1 = 2$, $1^2 + 2 = 3$, $2^2 + 3 = 7$. . . . So, to find the next two terms, you just need to continue the pattern: $7^2 + 16 = 65$ and $16^2 + 65 = 321$.

5 Find the sum of the first five terms of the series: $2^{n-1} + 1$.

The answer is 36. This one is an easy plug and chug. Simply plug in values 1 through 5 for n and simplify: $2^{1-1} + 1 = 2$, $2^{2-1} + 1 = 3$, $2^{3-1} + 1 = 5$, $2^{4-1} + 1 = 9$, $2^{5-1} + 1 = 17$. Now you just need to add the terms: $2 + 3 + 5 + 9 + 17 = 36$.

6 Find the sum of the first five terms of the series: $3^n + 2n$.

The answer is 393. Just like the last one, plug and chug away: $3^1 + 2(1) = 5$, $3^2 + 2(2) = 13$, $3^3 + 2(3) = 33$, $3^4 + 2(4) = 89$, $3^5 + 2(5) = 253$. Now just add 'em up: $5 + 13 + 33 + 89 + 253 = 393$.

7 Find the 50th term of the arithmetic sequence: –6, –1, 4, 9, . . .

The answer is $a_{50} = 239$. To start, find the formula for the nth term. For this, you need a_1, which is –6. Next, you need the common difference, d, found by subtracting two sequential terms: $a_2 - a_1 = -1 - (-6) = 5$. From here, simply plug these into the general formula and simplify: $a_n = -6 + (n-1)5 = -6 + 5n - 5 = 5n - 11$. Now that you have the general formula, you can find the 50th term by plugging in 50 for n: $a_{50} = 5(50) - 11 = 250 - 11 = 239$.

8 Find the general formula of an arithmetic sequence where $a_1 = -3$ and $a_{15} = 53$.

The answer is $a_n = 4n - 7$. Here, you have the first term, a_1, so you just need to find d. You can use a_{15} to find it. Simply plug in a_1, a_{15}, and $n = 15$ into the general formula and solve using algebra: $53 = -3 + (15-1)d$; $53 = -3 + 14d$; $56 = 14d$; $4 = d$. Now you can plug it in to find the general formula: $a_n = -3 + (n-1)4 = -3 + 4n - 4 = 4n - 7$.

9   Find the general formula of an arithmetic sequence where $a_5 = -5$ and $a_{20} = -35$.

The answer is $a_n = -2n + 5$. This one is a bit more complicated but completely doable! To start, recognize that you have two terms, which means you can create two equations. Then you have a system of equations (check out Chapter 13 for a refresher). Solve these to find your missing variables and you can write your general formula. Start by writing your two equations with your given values and simplify: $-5 = a_1 + (5 - 1)d = -5 = a_1 + 4d$ and $-35 = a_1 + (20 - 1)d = -35 = a_1 + 19d$. Next, use elimination to solve the system by multiplying the first equation by -1 and adding the two together to get $-30 = 15d$. Therefore, $d = -2$. Now, simply substitute this back into either equation to find a_1: $-5 = a_1 + 4(-2)$; $-5 = a_1 - 8$; $3 = a_1$. Finally, plug in a_1 and d to find the general formula: $a_n = 3 + (n - 1)(-2)$; $a_n = 3 + -2n + 2$; $a_n = -2n + 5$.

10   Find $\sum\limits_{n=1}^{5} \frac{1}{2}n + 2$.

The answer is $\frac{35}{2}$. To find the sum, you just have to use the arithmetic series formula. For this, you need k (which is 5), a_1, and a_k. Start by finding a_1: $\frac{1}{2} \cdot (1) + 2 = \frac{5}{2}$. Then find a_5: $\frac{1}{2} \cdot (5) + 2 = \frac{9}{2}$. Finally, plug these into the formula: $S_n = (\frac{5}{2})(\frac{5}{2} + \frac{9}{2}) = (\frac{5}{2})(\frac{14}{2}) = (\frac{5}{2})(7) = \frac{35}{2}$.

11   Find $\sum\limits_{n=4}^{10} 2n - 3$.

The answer is 77. Follow the same steps as the previous problem. However, notice that the lower limit is 4, so you need to start by finding a_4: $2(4) - 3 = 5$, and this is like your a_1. The number of terms from 4 to 10 is 7, so $k = 7$. You also need $a_{10} = 2(10) - 3 = 17$. Finally, just plug in the values: $S_n = (\frac{7}{2})(5 + 17) = (\frac{7}{2})(22) = 7(11) = 77$.

12   Write the arithmetic series: $2 + \frac{7}{3} + \frac{8}{3} + 3 + \frac{10}{3}$ in summation notation, and find the result.

The answer is $\sum\limits_{n=1}^{5} \frac{5}{3} + \frac{1}{3}n = \frac{40}{3}$. For summation notation, you need to find the general formula and know how many terms you're dealing with. In this case, you have five terms. Therefore, you know your upper limit, k, is 5. For the general formula, you need the first term $a_1 = 2$ and the common difference, d, which is found by subtracting two sequential terms: $a_2 - a_1 = \frac{7}{3} - 2 = \frac{1}{3}$. Plug these in and simplify to find your general formula: $a_n = 2 + (n - 1)\frac{1}{3} = 2 + \frac{1}{3} \cdot n - \frac{1}{3} = \frac{5}{3} + \frac{1}{3} \cdot n$. Then, plug the general formula into the summation notation and add the values given in the original problem to find the result: $\sum\limits_{n=1}^{5} \frac{5}{3} + \frac{1}{3}n = \frac{40}{3}$.

13   Find the 16th term of a geometric sequence given $a_1 = 5$ and $a_2 = -15$.

The answer is $a_{16} = -14,348,907$. To find the 16th term, you need to find the general formula. For that, you need a_1 and r. a_1 is given in the problem: 3. To find r, all you need to do is divide a_2 by a_1: $\frac{-15}{5} = -3$. Now you can simply plug these values into the formula: $a_{16} = 5 \cdot (-3)^{16-1} = 5 \cdot (-3)^{15} = -14,348,907$.

14   Find the 8th term of a geometric sequence given $a_2 = 6$ and $a_6 = 486$.

The answer is 4,374. This time you don't have the first term, so you have to set up a system of equations and simplify: $6 = a_1 \cdot r^{2-1}$; $6 = a_1 r^1$ and $486 = a_1 \cdot r^{6-1}$; $486 = a_1 r^5$. Isolate a_1 and use substitution to solve for r: $a_1 = \frac{6}{r}$; $486 = (\frac{6}{r})r^5$; $486 = 6r^4$; $81 = r^4$; $3 = r$. Then substitute r back into either equation to find a_1: $6 = a_1(3)$; $2 = a_1$. Now you can set up the general formula: $a_n = 2 \cdot 3^{n-1}$. Finally, to find the 8th term, plug in $n = 8$: $a_8 = 2 \cdot 3^{8-1} = 2 \cdot 3^7 = 2 \cdot 2,187 = 4,374$.

15   Find the sum $\sum\limits_{n=1}^{\infty} 3\left(-\frac{2}{3}\right)^{n-1}$.

The answer is $\frac{21}{8}$. Start by finding $a_1 = 4(-\frac{1}{2})^{1-1} = 4(-\frac{1}{2})^0 = 4 \cdot 1 = 4$. Because $r = -\frac{1}{2}$, you have everything you need to plug into partial sum formula:

$$S_6 = 4\left(\frac{1 - \left(-\frac{1}{2}\right)^6}{1 - \left(-\frac{1}{2}\right)}\right) = 4\left(\frac{1 - \frac{1}{64}}{1 + \frac{1}{2}}\right) = 4\left(\frac{\frac{63}{64}}{\frac{3}{2}}\right) = 4\frac{63}{64} \cdot \frac{2}{3} = \frac{21}{8}$$

16 Find the partial sum of the geometric series: $\frac{1}{6} + \frac{1}{3} + \frac{2}{3} + \ldots \frac{32}{3}$.

The answer is $\frac{127}{6}$. Here you have a_1 and you can find r by dividing a_2 by a_1: $\frac{1}{3} \div \frac{1}{6} = 2$. The trick here is that you need to find n. To do so, plug the last term into the general formula and use properties of exponents to solve for n: $\frac{32}{3} = \frac{1}{6} \cdot 2^{n-1}$; $64 = 2^{n-1}$; $2^6 = 2^{n-1}$; $6 = n - 1$; $7 = n$. Now that you have all the variables, plug them in to find the partial sum: $S_7 = \frac{1}{6}\left(\frac{1-2^7}{1-2}\right) = \frac{1}{6} \cdot \frac{-127}{-1} = \frac{127}{6}$.

17 Find the sum of the infinite geometric series: $\frac{2}{3} + \frac{1}{3} + \frac{1}{6} + \ldots$.

The answer is $\frac{4}{3}$. You need to start by finding r, which is a_2 divided by a_1: $\frac{1}{3} \div \frac{2}{3} = \frac{1}{2}$. a_1 is $\frac{2}{3}$. So all you need to do is plug these values into the appropriate formula and use what you know about fractions to simplify: $S = \frac{\frac{2}{3}}{1 - \frac{1}{2}} = \frac{\frac{2}{3}}{\frac{1}{2}} = \frac{2}{3} \cdot 2 = \frac{4}{3}$.

18 Find the sum: $\displaystyle\sum_{n-1}^{\infty} 3\left(-\frac{2}{3}\right)^{n-1}$.

The answer is $\frac{9}{5}$. Start by finding a_1 by plugging 1 into the general formula: $3(-\frac{2}{3})^{1-1} = 3(-\frac{2}{3})^0 = 3$. Next, notice that you have r, $-\frac{2}{3}$. From here, plug in these values to find the infinite sum: $S = \frac{3}{1 - \left(-\frac{2}{3}\right)} = \frac{3}{\frac{5}{3}} = 3 \cdot \frac{3}{5} = \frac{9}{5}$.

19 Find the 5th term of $(x + y)^{12}$.

The answer is $495x^8y^4$. All you have to do to find the 5th term is use the handy binomial theorem. In this case, n is 5, $a = x$, and $b = y$: $\binom{12}{4}x^{12-4}y^4$. Because the original binomial doesn't have any coefficients, the coefficient just comes from the combinations formula: $\frac{12!}{4!(12-4)!} = \frac{479,001,600}{24 \cdot 40,320} = 495$. Multiply the other terms by this and you get $495x^8y^4$.

20 Find the 8th term of $(2x - 3y)^{10}$.

The answer is $-2,099,520x^3y^7$. In this case, $a = 2x$, $b = -3y$, and n is 8. Plug these into the binomial theorem and simplify: $\binom{10}{7}(2x)^{10-7}(-3y)^7 = \frac{10!}{7!(10-7)!}(2x)^3(-3y)^7 = 120 \cdot 8x^3 \cdot -2,187y^7$ $= -2,099,520x^3y^7$.

21 Expand $(k - 4)^5$.

The answer is $k^5 - 20k^4 + 160k^3 - 640k^2 + 1280k + 1024$. To expand, simply replace a with k, b with -4, and n with 5 to get:

$$\binom{5}{0}(k)^5(-4)^0 + \binom{5}{1}(k)^4(-4)^1 + \binom{5}{2}(k)^3(-4)^2 + \binom{5}{3}(k)^2(-4)^3 + \binom{5}{4}(k)^1(-4)^4 + \binom{5}{5}(k)^0(-4)^5.$$

To simplify, start with the combinations formula for each term:

$$(1)(k)^5(-4)^0 + (5)(k)^4(-4)^1 + (10)(k)^3(-4)^2 + (10)(k)^2(-4)^3 + (5)(k)^1(-4)^4 + (1)(k)^0(-4)^5.$$

Next, raise the monomials to the specified powers:

$$(1)(k^5)(1) + (5)(k^4)(-4) + (10)(k^3)(16) + (10)(k^2)(-64) + (5)(k)(256) + (1)(1)(1024).$$

Last, combine like terms and simplify: $k^5 - 20k^4 + 160k^3 - 640k^2 + 1280k + 1024$.

22   Expand $(y + 4z)^6$.

The answer is $y^6 + 24y^5z + 240y^4z^2 + 1280y^3z^3 + 3,840y^2z^4 + 6,144yz^5 + 4,096z^6$. Here, replace a with y, b with $4z$, and n with 6 to get:

$$\binom{6}{0}(y)^6(4z)^0 + \binom{6}{1}(y)^5(4z)^1 + \binom{6}{2}(y)^4(4z)^2 + \binom{6}{3}(y)^3(4z)^3 + \binom{6}{4}(y)^2(4z)^4 + \binom{6}{5}(y)^1(4z)^5 + \binom{6}{6}(y)^0(4z)^6.$$

Then, to simplify this mess, start with the combinations formula for each term:

$$(1)(y)^6(4z)^0 + (6)(y)^5(4z)^1 + (15)(y)^4(4z)^2 + (20)(y)^3(4z)^3 + (15)(y)^2(4z)^4 + (6)(y)^1(4z)^5 + (1)(y)^6(4z)^6.$$

Raise the monomials to the specified powers:

$$(1)(y^6)(1) + (6)(y^5)(4z) + (15)(y^4)(16z^2) + (20)(y^3)(64z^3) + (15)(y^2)(256z^4) + (6)(y)(1024z^5) + (1)(1)(4096z^6).$$

Finally, combine like terms and simplify: $y^6 + 24y^5z + 240y^4z^2 + 1280y^3z^3 + 3,840y^2z^4 + 6,144yz^5 + 4,096z^6$.

Chapter 15

The Next Step Is Calculus

Your high school math career probably began with Algebra I, and then you repeated some of the information in Algebra II, and then again the same material presented itself in Pre-calculus (albeit a bit morphed and complicated since Algebra 1). The end of pre-calc is the beginning of calculus. Calculus teachers assume you've been paying attention during your math years, or they assume that you bought this book and are now a math genius, and that most of the material you've learned has stuck with you. Because of this, calc teachers move pretty quickly into new material.

Calculus is the study of change. Until now, all the information you use to solve problems has been constant, so your answers have always been constant. For example, up until calc, in a distance problem the rate at which a car is moving remains constant. The slope of a straight line is always a constant. The volume of a shape is always a constant. But in calculus, all these can move and grow and change. For example, the car can accelerate, decelerate, and accelerate again, all within the same problem, which changes the whole outcome. The line can now be a curve so that its slope changes over time. The shape that you're trying to find the volume of can get bigger or smaller, so that the volume changes over time.

All this change may send you screaming in terror to the nearest antianxiety salt lick, but chances are, if you've made it this far in your math career, you can handle it. So just in case your Pre-calc course ends with a preview of calc, we include the first couple of topics here.

Finding Limits: Graphically, Analytically, and Algebraically

If you haven't noticed by now, graphing functions has slowly become more complex and intricate. The more complicated the function is, the more complicated the graph tends to be. By now you've seen functions that are undefined at certain values; the graph has either a hole or a vertical asymptote, which affects your domain. The end of pre-calc (and the beginning of calc) looks at the *limit* of a function — what the function would do if it could.

In symbols, a limit is written as $\lim_{x \to n} f(x) = L$, which is read as "the limit of $f(x)$ as x approaches n is L." L is the limit that you're looking for. For the limit of a function to exist, the left limit and the right limit must be equal.

> ✔ A left limit starts at an x value that's less than the number n that x is approaching and gets closer and closer from the left. This is written as $\lim\limits_{x \to n^-} f(x)$.
>
> ✔ A right limit starts at an x value that's greater than the number n that x is approaching and gets closer and closer from the right. This is written as $\lim\limits_{x \to n^+} f(x)$.

When the left limit and right limit are the same, only then does the function have a limit. When $\lim\limits_{x \to n^-} f(x) = \lim\limits_{x \to n^+} f(x) = L$, then $\lim\limits_{x \to n} f(x) = L = L$.

You can find a limit in three different ways: graphically, analytically, and algebraically. We take a look at each one so you know how to handle them.

TIP

Before you try any of the techniques for finding a limit, always try plugging in the value that x is approaching into the function. If you get an answer, then that's also your limit. We also recommend only using the graphing method when you've been given the graph and asked to find the limit. The analytical method works for any function, and sometimes it's the only method you have in your pocket that will work. However, if the algebraic method works, then go with it — the analytical method is just too long and tedious.

Graphically

When you're given the graph of a function and asked to find the limit, just read the graph as x approaches the given value and see what the y value would have been (or was if the function is defined).

EXAMPLE

Q. In the given graph for f(x), find $\lim\limits_{x \to -3} f(x)$, $\lim\limits_{x \to 4} f(x)$, and $\lim\limits_{x \to 6} f(x)$.

A. $\lim\limits_{x \to -3} f(x) = \mathbf{1}$, $\lim\limits_{x \to 4} f(x) = \mathbf{-2}$, **and** $\lim\limits_{x \to 6} f(x) =$

doesn't exist. The function is defined at $x =$ -3 because you can see a dot there, so the limit of the function is the same as the function's value: f(x) = 1, so $\lim\limits_{x \to -3} f(x) = 1$.

The function isn't defined at $x = 4$ because the graph has a hole, but if you move along the graph from the left as x approaches 4, y also approaches a value of -2. Because this is the same from the right side, then $\lim\limits_{x \to 4} f(x) = -2$. The graph has an asymptote at $x = 6$, but the left limit is completely different from the right limit, so the limit as x approaches 6 doesn't exist. Some teachers and books write this as "DNE" (does not exist) because we see it frequently (and we're lazy as mathematicians).

1. In the given graph for g(x), find $\lim_{x \to -5} g(x)$, $\lim_{x \to -2} g(x)$, and $\lim_{x \to 1} g(x)$.

Solve It

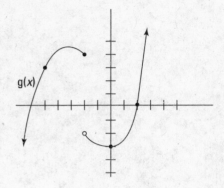

2. In the given graph for h(x), find $\lim_{x \to -3} h(x)$, $\lim_{x \to 5} h(x)$, and $\lim_{x \to 0} h(x)$.

Solve It

Analytically

"Analytically" means "systematically," and that's exactly how you find the limit using this method. You set up a chart, and the value that x is approaching goes in the middle of the top row. On the left you put values that get closer to the value that x is approaching, and you do the same thing on the right. The second row should be the y values when you plug the top row into the function. Hopefully, the bottom row values approach the same number from the left and right, and *voilà* — you have your limit!

Q. Find the limit of $\lim_{x \to 2} \dfrac{x^2 + 2x - 8}{x - 2}$.

A. $\lim_{x \to 2} \dfrac{x^2 + 2x - 8}{x - 2} = 6$. Notice that if you plug 2 into the function for x, you get 0 in the denominator. That means the function is undefined there. If the directions ask you to find this limit analytically, you'd set up a

chart. Here's the table that we set up. ***Note***: Yours doesn't have to look like this; there's no one way that works all the time in finding the limit analytically, but most of the tables we've seen are set up similarly. Try to get those x values really close to the value you're approaching — that's usually the best way to find the limit.

x	1	1.9	1.99	1.999	2	2.001	2.01	2.1	3
y	5	5.9	5.99	5.999	???	6.001	6.01	6.1	7

Notice that the y values from both the left and the right seem to be approaching 6; that's your limit.

3. Find the limit of $\lim\limits_{x \to -1} \dfrac{x^2 + 13x + 12}{x + 1}$.

Solve It

4. Find the limit of $\lim\limits_{x \to 2} \dfrac{3x^2 - 5x - 2}{x - 2}$.

Solve It

Algebraically

To find a limit algebraically, you can use four different techniques: plugging in, factoring, rationalizing the numerator, and finding the lowest common denominator. Always start by plugging the given number into the function just to see whether it works. If the answer is undefined, then move on to one of the other three techniques — each one depending on the given function.

Plugging in

This first technique asks you to substitute the given value into the function. If you get an undefined value, like 0 in the denominator of the fraction, try something else. But when it works, this technique is the quickest way to find a limit. We like short methods, and we hope you do, too!

Factoring

When the function is a rational function with polynomials in the numerator and the denominator, try factoring them. If you've forgotten how to factor a polynomial and need a refresher, see Chapter 4. Chances are, some factors will cancel from both the numerator and the denominator. You can then substitute the given value into the cancelled version and usually get an answer that's also your limit.

When you follow the steps of factoring the rational function on the top and bottom, canceling, and plugging in the given value, if you still get an undefined function, then the limit does not exist (DNE).

Rationalizing the numerator

When you see square roots in the rational function in the numerator and plugging in doesn't work, always try to rationalize the numerator. That's right — you multiply by the conjugate of the numerator on both the top and bottom of the fraction. When you do, you usually see a bunch of terms cancel, and the function will simplify down to a point where you can plug in the given value and find the limit.

Finding the lowest common denominator

When the rational function is a *complex rational function* (fractions inside of the fraction), find the common denominators and add or subtract terms, then cancel and simplify. You can then plug in the given value to find the limit.

Q. Find the limit of $\lim\limits_{x \to 0} \dfrac{\dfrac{1}{x+5} - \dfrac{1}{5}}{x}$.

A. $\lim\limits_{x \to 0} \dfrac{\dfrac{1}{x+5} - \dfrac{1}{5}}{x} = -\dfrac{1}{25}$. Find the common denominator of the fractions on the top first:

$\lim\limits_{x \to 0} \dfrac{\dfrac{5 \cdot 1}{5(x+5)} - \dfrac{1(x+5)}{5(x+5)}}{x} = \dfrac{\dfrac{5}{5(x+5)} - \dfrac{x+5}{5(x+5)}}{x}$. Subtract and then simplify the top now that you have

a common denominator: $\dfrac{\dfrac{5-x-5}{5(x+5)}}{x} = \dfrac{\dfrac{-x}{5(x+5)}}{x} = \dfrac{-x}{5(x+5)} \div x = \dfrac{-x}{5(x+5)} \cdot \dfrac{1}{x} = \dfrac{-1}{5(x+5)}$. Notice that

you can now plug 0 into this function and get the limit: $-\dfrac{1}{25}$.

---

5. Find the limit of $\lim\limits_{x \to 2} \dfrac{3x^2 - 5x - 2}{x - 2}$ algebraically.

Solve It

6. Find the limit of $\lim\limits_{x \to 5} \dfrac{\sqrt{x-1} + 2}{x - 5}$.

Solve It

Knowing Your Limit Laws

Calculus also provides you with a few limit laws that help you find the limits of combined functions: added, subtracted, multiplied, divided, and even raised to powers. If you can find the limit of each individual function, you can find the limit of the combined function as well.

If $\lim\limits_{x \to n} f(x) = L$ and $\lim\limits_{x \to n} g(x) = M$, the limit laws are:

- **Addition law:** $\lim\limits_{x \to n}\big(f(x) + g(x)\big) = L + M$
- **Subtraction law:** $\lim\limits_{x \to n}\big(f(x) - g(x)\big) = L - M$
- **Multiplication law:** $\lim\limits_{x \to n}\big(f(x) \cdot g(x)\big) = L \cdot M$
- **Division law:** $\lim\limits_{x \to n}\left(\dfrac{f(x)}{g(x)}\right) = \dfrac{L}{M}$
- **Power law:** $\lim\limits_{x \to n}\big(f(x)\big)^{p} = L^{p}$

For this whole section, use the following to answer the questions:

$$\lim_{x \to 1} f(x) = -5 \qquad \lim_{x \to 1} g(x) = 2 \qquad \lim_{x \to 1} h(x) = 0$$

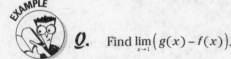

Q. Find $\lim\limits_{x \to 1}\big(g(x) - f(x)\big)$.

A. $\lim\limits_{x \to 1}\big(g(x) - f(x)\big) = $ **7**. The limit of $g(x)$ is 2 and the limit of $f(x)$ is –5. To find $\lim\limits_{x \to 1}\big(g(x) - f(x)\big)$, use the subtraction law: $2 - (-5) = 7$. It really is that easy!

7. Find $\lim\limits_{x \to 1}\dfrac{f(x)}{g(x)}$.

8. Find $\lim\limits_{x \to 1} f(x)^{2} - 2h(x) + \dfrac{1}{g(x)}$.

Solve It

9. Find $\lim\limits_{x \to 1} \dfrac{\sqrt{g(x)} - f(x)}{5g(x)}$.

Solve It

10. Find $\lim\limits_{x \to 1} \dfrac{f(x)}{h(x)}$.

Solve It

Determining Continuity

The word "continuity" in math means the same thing as it does in your everyday life. Something that's continuous has a stability or a permanence to it . . . it just doesn't ever stop. In pre-calc, you've seen functions that have holes in their graph, jumps in their graph, or asymptotes — just to name a few. A graph that doesn't have holes, jumps, or asymptotes keeps going forever, and we call that function *continuous*.

Polynomial functions, exponential functions, and logarithmic functions are continuous at every point. If you're ever asked to determine the continuity of one of these types of functions, don't bother — the answer is that it's always continuous.

We usually look at specific values in the domain to determine continuity instead of looking at the entire function. Even discontinuous functions are discontinuous only at certain places. The discontinuity at a certain x value in any function is either always *removable* (a hole in the graph) or *nonremovable* (an asymptote). It all depends on the factored version of the polynomial in the denominator. If a factor cancels, the discontinuity there is removable. If the factor doesn't cancel, the discontinuity is nonremovable.

Three things must be true for a function to be continuous at $x = c$:

- **$f(c)$ must be defined.** When you plug c into the function, you must get a value out again. For example, getting 0 in the denominator is unacceptable and therefore a discontinuity.

- **The limit of the function as x approaches c must exist.** The left and right limits must be the same. If they aren't, the function is discontinuous there.

- **The function's value and the limit must be the same.** $f(c) = \lim\limits_{x \to c} f(x)$. If the value of the function is one thing and the limit is something different, that's not good; the function is discontinuous there.

Here's the graph of a function where each one of the above situations fails:

At f(4) the graph has a hole. The function is undefined there, and therefore discontinuous at $x = 4$.

At f(–3) the function jumps. The limit as x approaches –3 from the left is 4, and from the right the limit is 1, so the limit doesn't exist, and the function is discontinuous.

At f(0) the function is defined at one point: f(0) = –4, but the limit as x approaches 0 from the left and from the right is –2. These two values must be the same for the function to be continuous.

One point where the function is continuous is at $x = 2$. $f(2) = \lim_{x \to 2} f(x) = 2$.

Q. Is $f(x) = \dfrac{3}{x+2}$ continuous at $x = 3$?

A. **f(x) is continuous at x = 3.** First, check to make sure the function is defined at $x = 3$. When you plug 3 in for x you get f(3) = ⅗. Next, make sure the limit is defined. Because you can plug in the value on the last step and get a value, you can also plug 3 in and get ⅗ for the limit. Because these two values are the same, the function is continuous at $x = 3$.

Q. Explain why $f(x) = \dfrac{3}{x+2}$ is not continuous at $x = –2$. Is this discontinuity removable or nonremovable?

A. **The function is not defined at f(–2).** This rational function doesn't factor, and therefore, nothing cancels and the discontinuity that exists is nonremovable.

11. Determine whether $g(x) = \dfrac{x^2 - 4x - 5}{x^2 - 2x - 15}$ is continuous at $x = 5$.

Solve It

12. Is $g(x) = \dfrac{x^2 - 4x - 5}{x^2 - 2x - 15}$ continuous at $x = 0$?

Solve It

13. Is $g(x) = \dfrac{x^2 - 4x - 5}{x^2 - 2x - 15}$ continuous at $x = -3$?

Solve It

14. Determine whether $h(x) = \begin{cases} 3x - 1 \text{ if } x \le -2 \\ x^2 - 11 \text{ if } x > -2 \end{cases}$ is continuous at $x = -2$.

Solve It

15. Determine whether $p(x) = \begin{cases} \dfrac{1}{2}x - 3 \text{ if } x \le 1 \\ 4x + 3 \text{ if } x > 1 \end{cases}$ is continuous at $x = 1$.

Solve It

16. Determine all numbers at which $q(x) =$
$x^2 - 3 \text{ if } x < -2$
$x + 3 \text{ if } -2 \le x <$ is continuous.
$x^2 - 3x \text{ if } x \ge 1$

Solve It

Answers to Problems on Calculus

1 In the given graph for g(x), find $\lim\limits_{x \to -5} g(x)$, $\lim\limits_{x \to -2} g(x)$, and $\lim\limits_{x \to 1} g(x)$. **$\lim\limits_{x \to -5} g(x) = 3$, $\lim\limits_{x \to -2} g(x)$ DNE, and**

$\lim\limits_{x \to 1} g(x) \approx -2.2.$

Looking at the graph for g(x), when $x = -5$, g(x) = 3. When x approaches -2 from the left the limit is 4, but when x approaches -2 from the right the limit is -2. Because these two values aren't the same, $\lim\limits_{x \to -2} g(x)$ does not exist (DNE). And when x approaches 1, the graph has a y value

somewhere between -2 and -3. Because this isn't an exact science, just do your best to approximate the limit. It looks to us like it's about -2.2.

2 In the given graph for h(x), find $\lim\limits_{x \to -3} h(x)$, $\lim\limits_{x \to 5} h(x)$, and $\lim\limits_{x \to 0} h(x)$. **$\lim\limits_{x \to -3} h(x) = \infty$, $\lim\limits_{x \to 5} h(x) = 2$, and**

$\lim\limits_{x \to 0} h(x) \approx -2.5.$

This answer also comes directly from the given graph. Trace your fingers along the graph as x gets closer and closer to -2 from the left and notice that the line keeps going up. Also notice that as x approaches this same value from the right, the graph is also going up. Like, forever and ever up. That's why $\lim\limits_{x \to -3} h(x) = \infty$. Meanwhile, as x approaches 5, h(x) approaches 2, and as

x approaches 0, it looks like f(x) is awful close to -2.5.

3 Find the limit $\lim\limits_{x \to -1} \dfrac{x^2 + 13x + 12}{x + 1}$. **The answer is 11.**

See our chart for the analytical evaluation of this limit:

x	−2	−1.1	−1.01	−1.001	−1	−0.999	−0.99	−0.9	0
y	10	10.9	10.99	10.999	???	11.001	11.01	11.1	12

By looking at the y values in the second row, it looks like from both the left and the right that y is approaching 11.

4 Find the limit $\lim\limits_{x \to 2} \dfrac{3x^2 - 5x - 2}{x - 2}$. **The answer is 7.**

Here's the chart we used for this limit:

x	1	1.9	1.99	1.999	2	2.001	2.01	2.1	3
y	4	6.7	6.97	6.997	???	7.003	7.03	7.3	10

Looking at this one, we call it 7. What do you think?

5 Find the limit of $\lim\limits_{x \to 2} \dfrac{3x^2 - 5x - 2}{x - 2}$ algebraically. **The answer is 7.**

Plugging 2 into the function gives you a 0 in the denominator, so you must try another technique. This rational function has a numerator that factors. (You did the limit analytically in

problem 4.) When you factor it, you should get $\dfrac{(3x + 1)(x - 2)}{x - 2}$. This reduces to $3x + 1$, which

gives you a function that you *can* plug 2 into: 3(2) + 1 = 6 + 1 = 7.

6 Find the limit of $\lim\limits_{x \to 5} \dfrac{\sqrt{x - 1} + 2}{x - 5}$. **The answer is $\dfrac{1}{3}$.**

Substitute 5 into this equation and you also get 0 in the denominator. Noticing that the numerator has a square root, you should find yourself thinking something along the lines of, "Perhaps I should multiply by the conjugate to rationalize the numerator." If you did think that or something close to it, give yourself a huge pat on the back.

Here's how to multiply by the conjugate: $\lim\limits_{x\to 5} \dfrac{\sqrt{x-1}+2}{x-5} \cdot \dfrac{\sqrt{x-1}-2}{\sqrt{x-1}-2}$. FOIL out the numerators and

watch the square roots disappear. However, don't multiply out the denominators — it'll cancel

easier if you don't: $\lim\limits_{x\to 5} \dfrac{x-1-2\sqrt{x-1}+2\sqrt{x-1}-4}{\left(x-5\right)\left(\sqrt{x-1}+2\right)} = \lim\limits_{x\to 5} \dfrac{x-1-4}{\left(x-5\right)\left(\sqrt{x-1}+2\right)} = \lim\limits_{x\to 5} \dfrac{x-5}{\left(x-5\right)\left(\sqrt{x-1}+2\right)}$.

The numerator and the factor on the left on the bottom both cancel and give you

$\lim\limits_{x\to 5} \dfrac{\cancel{x-5}}{\left(\cancel{x-5}\right)\left(\sqrt{x-1}+2\right)} = \lim\limits_{x\to 5} \dfrac{1}{\sqrt{x-1}+2}$. Now when you plug 2 in, you find that the limit is $\dfrac{1}{3}$.

7 Find $\lim\limits_{x\to 1} \dfrac{f(x)}{g(x)}$. **The answer is** $-\dfrac{5}{2}$.

Because you know both limits, to find the limit of their quotient, divide their limits as well:
$\lim\limits_{x\to 1} \dfrac{f(x)}{g(x)} = -\dfrac{5}{2}$.

8 Find $\lim\limits_{x\to 1} f(x)^2 - 2h(x) + \dfrac{1}{g(x)}$. **The answer is** $\dfrac{51}{2}$.

Plug in the information that you know based on the given limits: $\lim\limits_{x\to 1} f(x)^2 - 2h(x) + \dfrac{1}{g(x)} = (-5)^2$

$-2\left(0\right) + \dfrac{1}{2} = 25 - 0 + \dfrac{1}{2} = \dfrac{51}{2}$.

9 Find $\lim\limits_{x\to 1} \dfrac{\sqrt{g(x)} - f(x)}{5g(x)}$. **The answer is** $\dfrac{5+\sqrt{2}}{10}$.

Plug and chug away: $\lim\limits_{x\to 1} \dfrac{\sqrt{g(x)} - f(x)}{5g(x)} = \dfrac{\sqrt{2} - \left(-5\right)}{5\left(2\right)} = \dfrac{5+\sqrt{2}}{10}$.

10 Find $\lim\limits_{x\to 1} \dfrac{f(x)}{h(x)}$. $\lim\limits_{x\to 1} \dfrac{f(x)}{h(x)}$ **is undefined.**

This time, putting the limit of $h(x)$ in the denominator also puts 0 in the denominator. That's an undefined limit.

11 Determine whether $g(x) = \dfrac{x^2 - 4x - 5}{x^2 - 2x - 15}$ is continuous at $x = 5$. **$x = 5$ is a removable discontinuity.**

Factor the given equation first: $\dfrac{x^2 - 4x - 5}{x^2 - 2x - 15} = \dfrac{\left(x-5\right)\left(x+1\right)}{\left(x-5\right)\left(x+3\right)}$. Cancel to get $\dfrac{x+1}{x+3}$. Notice that

when you plug 5 into this simplified expression, you do get an answer of $\dfrac{6}{8}$, or $\dfrac{3}{4}$. But this isn't

the original, given equation. The graph is going to look and act like $\dfrac{x+1}{x+3}$, but because the origi-

nal equation is $\dfrac{x^2 - 4x - 5}{x^2 - 2x - 15}$ or $\dfrac{\left(x-5\right)\left(x+1\right)}{\left(x-5\right)\left(x+3\right)}$, there is still going to be a hole in the graph (try

plugging 5 into either of them and see what happens). This is why 5 is a removable discontinuity.

12 Is $g(x) = \dfrac{x^2 - 4x - 5}{x^2 - 2x - 15}$ continuous at $x = 0$? **The answer is yes, $x = 0$ is continuous.**

You can simply plug 0 into this function and get a value out: $\dfrac{0^2 - 4\left(0\right) - 5}{0^2 - 2\left(0\right) - 15} = \dfrac{-5}{-15} = \dfrac{1}{3}$. This makes the function continuous.

13 Is $g(x) = \dfrac{x^2 - 4x - 5}{x^2 - 2x - 15}$ continuous at $x = -3$? **$x = -3$ is a nonremovable discontinuity.**

Plugging -3 into the original function $\dfrac{x^2 - 4x - 5}{x^2 - 2x - 15}$ gives you 0 in the denominator, so you know

right away that it's discontinuous. When you factor and simplify to $\dfrac{x+1}{x+3}$, you *still* get 0 in the
denominator, so the discontinuity is nonremovable.

14 Determine whether $h(x) = \begin{cases} 3x - 1 \text{ if } x \le -2 \\ x^2 - 11 \text{ if } x > -2 \end{cases}$ is continuous at $x = -2$. **The function is continuous at $x = -2$.**

If you don't know how to deal with piece-wise functions like this, get a refresher from Chapter 3.

First, look at $f(-2) = 3(-2) - 1 = -6 - 1 = -7$. The function exists.

Now, look at $\lim\limits_{x \to -2^-} f(x) = 3(-2) - 1 = -6 - 1 = -7$. Next, look at $\lim\limits_{x \to -2^+} f(x) = (-2)^2 - 11 = 4 - 11 = -7$.
Because the left limit matches the right limit, the function has a limit as x approaches -2.

Lastly, because the function value matches the limit value, the function is continuous at $x = -2$.

15 Determine whether $p(x) = \begin{cases} \dfrac{1}{2}x - 3 \text{ if } x \le 1 \\ 4x + 3 \text{ if } x > 1 \end{cases}$ is continuous at $x = 1$. **The function isn't continuous at $x = 1$.**

$p(1) = \dfrac{1}{2}(1) - 3 = \dfrac{1}{2} - 3 = \dfrac{-5}{2}$. The function exists.

$\lim\limits_{x \to 1^-} p(x) = \dfrac{1}{2}(1) - 3 = \dfrac{1}{2} - 3 = \dfrac{-5}{2}$, but $\lim\limits_{x \to 1^+} p(x) = 4(1) + 3 = 4 + 3 = 7$. These two values aren't equivalent, so there is no limit and the function is discontinuous at $x = 1$.

16 Determine all numbers at which $q(x) = \begin{cases} x^2 - 3 \text{ if } x < -2 \\ x + 3 \text{ if } -2 \le x < 1 \\ x^2 - 3x \text{ if } x \ge 1 \end{cases}$ is continuous. **The function is continuous everywhere except $x = 1$.** In interval notation, this is written as $\left(-\infty, 1\right) \cup \left(1, \infty\right)$.

The only places that piece-wise functions have potential discontinuities are where the function may break into pieces — where the interval begins or ends.

$\lim\limits_{x \to -2^-} q(x) = (-2)^2 - 3 = 4 - 3 = 1$, and $\lim\limits_{x \to -2^+} q(x) = -2 + 3 = 1$. $\lim\limits_{x \to -2^-} q(x) = \lim\limits_{x \to -2^+} q(x)$, so the graph is continuous there.

$\lim\limits_{x \to 1^-} q(x) = 1 + 3 = 4$, but $\lim\limits_{x \to 1^+} q(x) = 1^2 - 3(1) = 1 - 3 = -2$. $\lim\limits_{x \to 1^-} q(x)$ doesn't equal $\lim\limits_{x \to 1^+} q(x)$, so the graph is discontinuous there.

Part V
The Part of Tens

The 5th Wave   By Rich Tennant

"Remember, basil is a polynomial and we'll be using two tablespoons of epsilon. But first, let's test to see if the vinegar is greater than or equal to the oil."

In this part . . .

This part has a summary of the parent graphs we cover in Chapter 3, including how to transform them. Think of it as a quick guide to all the topics we cover regarding graphing and transforming parent functions. This part also includes a chapter on the mistakes we commonly see in pre-calc and how to avoid them (please avoid them!).

Chapter 16

Ten Uses for Parent Graphs

A picture is worth a thousand words, and graphing is just math in pictures! These pictures can give you important information about the characteristics of a function. The most common graphs are called parent graphs. These graphs are in their original, unshifted form. Any parent graph can be stretched, shrunk, shifted, or flipped. They're extremely useful because you can use them to graph a more complicated version of the same function using transformations (see Chapter 3). That way if you're given a complex function (say a crazy-looking quadratic), you automatically have a basic idea of what the graph will look like without having to plug in a whole bunch of numbers first. Essentially, by knowing what the parent looks like, you get a good idea about the kids . . . the apple doesn't fall far from the tree, right? Now, let's check out those family pictures!

Squaring Up with Quadratics

The basic quadratic is simplicity itself: $y = x^2$. Its graph is a parabola with a vertex at the origin, reflected over the y-axis (see Figure 16-1). You can find out more about graphing quadratics in Chapters 3 and 12.

Figure 16-1:
The graph of
the function
$y = x^2$
passes
through the
origin and is
symmetric
across the
y-axis.

Cueing Up for Cubics

The parent graph of the cubic function, $y = x^3$, also passes through the origin. This graph is symmetric over the origin (see Figure 16-2). We cover cubics in Chapter 3.

Figure 16-2:
The graph of the parent function $y = x^3$.

Rooting for Square Roots and Cube Roots

A square root graph looks like a parabola that has been rotated clockwise 90° and cut in half. It's cut in half (only positive) because you can't take the square root of a negative number. The parent graph is pictured in Figure 16-3.

Figure 16-3:
The parent graph of $y = \sqrt{x}$ starts at the origin and curves up to the right.

Cube root functions are the inverse of cubic functions, so their graphs are going to reflect that. The parent graph of a cube root function passes through the origin and is symmetric over it, as shown in Figure 16-4.

Figure 16-4:
The parent graph of $y = \sqrt[3]{x}$ has a point of inflection (bend) at the origin.

Graphing Absolutely Fabulous Absolute Value Functions

Because the absolute value function turns all inputs into non-negative values (0 or positive), the parent graph is only above the *x*-axis. Figure 16-5 shows the parent graph in its characteristic V shape.

Figure 16-5:
The graph of the function $y = |x|$ has a vertex at the origin and is symmetric over the *y*-axis.

Flipping over Rational Functions

In Chapter 3, we take you through the steps for graphing rational functions. These involve finding asymptotes, intercepts, and key points. Because these functions don't really have parent graphs per se, we thought we'd show you an example of the most basic rational function: $y = \frac{1}{x}$ (see Figure 16-6). To see more, flip back to Chapter 3.

Figure 16-6: A graph of the rational function $y = \frac{1}{x}$ with its asymptotes.

Exploring Exponential Graphs and Logarithmic Graphs

The parent graph of an exponential function is $y = b^x$ where b is the base. Because b has to be some number to graph, we thought we'd show you the graph of $y = e^x$. This graph passes through the point $(0, 1)$ and a horizontal asymptote of the x-axis, as shown in Figure 16-7. We cover exponential graphs in Chapter 5.

Figure 16-7: The exponential graph of $y = e^x$ increases without bound to the right.

The inverse (see Chapter 3) of an exponential function is a logarithmic function. So we're going to show you the inverse of the graph of $y = e^x$, which is the graph of $y = \log_e x$, which is known as the natural log, or $y = \ln x$ (see Chapter 5). This graph passes through the point (1, 0) and a vertical asymptote of the y-axis, as shown in Figure 16-8. We also cover logarithmic graphs in Chapter 5.

Figure 16-8:
The loga-
rithmic
graph of
$y = \log_e x$
increases
slowly to the
right.

Seeing the Sine and Cosine

A sine graph looks like a wave. The parent graph passes through the origin and has an amplitude of 1. The period is 2π, which means that the wave repeats itself every 2π. Figure 16-9 shows one full period of the parent sine graph.

Figure 16-9:
The parent
graph of
sine:
$y = \sin\theta$.

Like sine, the graph of cosine is a wave. This parent graph passes through the point (0, 1) and also has an amplitude of 1 and a period of 2π. You can see the parent graph in Figure 16-10. For more information about graphing sine and cosine, turn to Chapter 7.

Figure 16-10:
The parent
graph of
cosine:
$y = \cos\theta$.

Covering Cosecant and Secant

Remember reciprocals? Well, cosecant is the reciprocal of sine, so the graph of cosecant reflects that. We've lightly sketched the parent sine graph along with the parent cosecant graph in Figure 16-11 so you can see the relationship. Specific graphing information can be found in Chapter 7.

Figure 16-11: The graph of the parent cosecant function: $y = \csc\theta$ with a light sketch of $y = \sin\theta$.

Again, like cosecant, secant is the reciprocal of cosine, so the graph of secant is similar to the graph of cosine. To picture this, we've lightly drawn the graph of cosine along with the parent graph of secant in Figure 16-12.

Figure 16-12: The graph of the parent secant function: $y = \sec\theta$ with a light sketch of $y = \cos\theta$.

Tripping over Tangent and Cotangent

One repeating pattern of the graph of tangent is its asymptotes, where the function is undefined. Like other trig graphs, a tangent graph has a period where it repeats itself. In this case, it's π. In Figure 16-13, we show you one period of the parent tangent graph. For more information about graphing tangents, turn to Chapter 7.

Figure 16-13:
One period
of the
parent
tangent
function:
$y = \tan\theta$.

Like tangents, the parent graph of cotangent has asymptotes at regular intervals. Also like tangents, the period of cotangent is π. In Figure 16-14, we show you one period of the parent cotangent graph. You can get more information about graphing cotangents in Chapter 7.

Figure 16-14:
The graph of
the parent
function
$y = \cot\theta$.

Chapter 17

Ten Pitfalls to Pass Up in Pre-Calc

*O*ne of our favorite video games from childhood (we know this dates us, but . . .) featured a little guy with a square head running through the square jungle, swinging on square vines, and jumping over square alligators in square swamps. He was avoiding the pitfalls of the jungle. Consider this chapter the vine you can use to jump over the pitfalls that normally trip up the pre-calc student. And you don't have to be a video game geek to understand this chapter!

Going Out of Order (Of Operations)

Operations in an expression or an equation aren't meant to be done from left to right. For example, $3 - 7(x - 2)$ doesn't equal $-4(x - 2)$ or $-4x + 8$. You're supposed to do multiplication first, which means distributing the -7 first: $3 - 7x + 14$. Now combine like terms to get $-7x + 17$.

Remember your order of operations (PEMDAS) all the time, every time:

> **P**arentheses (and other grouping devices)
>
> **E**xponents
>
> **M**ultiplication and **D**ivision, from left to right in order as you find them
>
> **A**ddition and **S**ubtraction, also from left to right

To further review the order of operations, see Chapter 1.

FOILed Again! FOILing Binomials Incorrectly

When multiplying binomials, always remember to multiply them in the correct order. You remember FOIL — First, Outside, Inside, Last. This includes squaring any binomial. The biggest mistake we see in these situations is something like: $(x - 4)^2 = x^2 + 16$. That's forgetting a whole lot of multiplying, though. It should look like this: $(x - 4)^2 = (x - 4)(x - 4) = x^2 - 4x - 4x + 16 = x^2 - 8x + 16$.

Gettin' Down with Denominators

Don't fall for a big trap and break a fraction up incorrectly. $\frac{5}{4x-3}$ doesn't equal $\frac{5}{4x}+\frac{5}{3}$.

If you don't believe us, pick a value for x and plug it into both equations and see if you get the same answer twice. You won't, because it doesn't work. You're doing the order

of operations wrong if you change $\frac{6}{3x-12}$ to $\frac{2}{x-4}$ (see the section "Going Out of Order

(Of Operations)," earlier in this chapter). The division bar is a grouping symbol, and you have to simplify the numerator and denominator separately before doing the divi-

sion. $\frac{6}{3x-12}$ is simply $\frac{6}{3x-12}$, it doesn't simplify. However, you may come across a

case where a fraction does simplify. For example, $\frac{3}{x+1} \div \frac{2}{x-2}$ simplifies to $\frac{3}{x+1} \cdot \frac{2}{x-2}$

because 3 goes into every term.

Combining Terms That Can't Be Combined

Yet another mistake we see frequently is students combining terms that aren't meant to be combined.

$4x - 1$ suddenly becomes $3x$, which it's not. $4x - 1$ is simplified, meaning that it's an equation that doesn't contain any like terms. $3a^4b^5 + 2a^5b^4$ is also simplified. Those exponents are close, but close only counts in horseshoes and hand grenades. When counting in the real world (as opposed to the algebra one you probably feel stuck in now), you can't combine apples and bananas. Four apples plus three bananas is still four apples and three bananas. It's the same in algebra: $4a + 3b$ is simplified.

Forgetting to Flip the Fraction

When dealing with complex fractions, don't take all the rules that you've learned and

throw them out the window. $\dfrac{\frac{3}{x+1}}{\frac{2}{x-2}}$ doesn't become $\frac{3}{x+1} \cdot \frac{2}{x-2}$. If you think it does,

you're probably forgetting that a division bar is division.

$\dfrac{\frac{3}{x+1}}{\frac{2}{x-2}} = \frac{3}{x+1} \div \frac{2}{x-2}$. To divide a fraction, you must multiply by its reciprocal.

$\frac{3}{x+1} \div \frac{2}{x-2} = \frac{3}{x+1} \cdot \frac{x-2}{2}$.

Losing the Negative (Sign)

We know that in life you're not supposed to be negative, but in math, don't throw away a negative sign — especially when subtracting polynomials.

$(4x^3 - 6x + 3) - (3x^3 - 2x + 4)$ isn't the same thing as $4x^3 - 6x + 3 - 3x^3 - 2x + 4$. If you did it that way, you didn't subtract the whole second polynomial, only its first term. The right way to do it is $4x^3 - 6x + 3 - 3x^3 + 2x - 4$, which simplifies to $x^3 - 4x - 1$.

Similarly, when subtracting rational functions, take care of that negative sign.

$\dfrac{3x+5}{x-2} - \dfrac{x-6}{x-2} \neq \dfrac{3x+5-x-6}{x-2}$. What happened? You forgot to subtract the whole

second polynomial on the top. Instead, you should've done this:

$$\frac{3x+5}{x-2} - \frac{x-6}{x-2} = \frac{(3x+5)-(x-6)}{x-2} = \frac{3x+5-x+6}{x-2} = \frac{2x+11}{x-2}.$$

Oversimplifying Roots

When it comes to roots, we've seen all kinds of errors. For instance, $\sqrt{3}$ suddenly

becomes 3 in a problem and loses the root altogether. Or $\dfrac{3x+5}{x-2} - \dfrac{x-6}{x-2} \neq \dfrac{3x+5-x-6}{x-2}$

becomes $\sqrt{2}$, losing the index on the root.

Don't add or subtract roots that aren't like terms, either. $\sqrt{7} + \sqrt{3}$ isn't $\sqrt{10}$, now or

ever. They're not like terms, so you can't add them. $\sqrt{7} + \sqrt{3}$ is it — it's done.

Avoiding Exponent Errors

When multiplying monomials, don't multiply the exponents. We've seen students who've been dealing with exponents for a long time suddenly do something like this: $x^4 \cdot x^3 = x^{12}$. Also, when dealing with a power over a product, you must apply the power to everything. $(2x^5y)^3$ isn't $2x^5y^3$ or $2x^{15}y^3$. You must raise everything inside the parentheses to the third power, so the answer should be $8x^{15}y^3$.

Watch out when dealing with negatives and exponents in some calculators. -4^2 and $(-4)^2$ represent -16 and 16 respectively, so be sure you know which configuration you're looking for when you punch it into your calculator.

Canceling Too Quickly

You can cancel terms when adding or subtracting if you have two terms that are exact opposites of each other. In multiplication and division, you can cancel terms if one common factor divides into all terms.

Here are the most common canceling mistakes we see:

- **Canceling constants.** If you see the rational expression $\dfrac{15x-2}{5}$, it doesn't equal $3x - 2$. The 5 in the denominator has to divide into both terms on the top.

- **Canceling variables.** For the same reason as the preceding bullet, $\dfrac{4x^2 - 3x + 2}{x}$ isn't $4x - 3x + 2$ because the denominator doesn't divide into everything on the top, only the first term.

- **Canceling everything.** To keep you aware, $\dfrac{4x^2 - 3x + 2}{2x^2 - 6x + 8}$ isn't

$$\frac{\cancel{4}^{2}\,\cancel{x^2} - \cancel{3}\,\cancel{x} + 2}{\cancel{2}\,x^2 - \cancel{6}_{2}\,\cancel{x} + \cancel{8}_{4}} = \frac{2}{2+4} = \frac{2}{6} = \frac{1}{3}$$, which would be creating a constant out of two

polynomials dividing.

Dealing with Distribution

When a polynomial is multiplied by a monomial, the process is known as *distribution*. Think of it like delivering the newspaper to every house on the block.

We've also seen students who don't distribute to every term, especially when the polynomial gets long.

$3(2x^5 - 6x^4 + 3x^3 - x^2 + 7x - 1)$ isn't equal to

$6x^5 - 6x^4 + 3x^3 - x^2 + 7x - 1$,

or $6x^5 - 18x^4 + 3x^3 - x^2 + 7x - 1$,

or even $6x^5 - 18x^4 + 9x^3 - x^2 + 7x - 1$,

or anything other than $6x^5 - 18x^4 + 9x^3 - 3x^2 + 21x - 3$.

Index

• D •

Notes

BUSINESS, CAREERS & PERSONAL FINANCE

Accounting For Dummies, 4th Edition*
978-0-470-24600-9

Bookkeeping Workbook For Dummies†
978-0-470-16983-4

Commodities For Dummies
978-0-470-04928-0

Doing Business in China For Dummies
978-0-470-04929-7

E-Mail Marketing For Dummies
978-0-470-19087-6

Job Interviews For Dummies, 3rd Edition*†
978-0-470-17748-8

Personal Finance Workbook For Dummies*†
978-0-470-09933-9

Real Estate License Exams For Dummies
978-0-7645-7623-2

Six Sigma For Dummies
978-0-7645-6798-8

Small Business Kit For Dummies, 2nd Edition*†
978-0-7645-5984-6

Telephone Sales For Dummies
978-0-470-16836-3

BUSINESS PRODUCTIVITY & MICROSOFT OFFICE

Access 2007 For Dummies
978-0-470-03649-5

Excel 2007 For Dummies
978-0-470-03737-9

Office 2007 For Dummies
978-0-470-00923-9

Outlook 2007 For Dummies
978-0-470-03830-7

PowerPoint 2007 For Dummies
978-0-470-04059-1

Project 2007 For Dummies
978-0-470-03651-8

QuickBooks 2008 For Dummies
978-0-470-18470-7

Quicken 2008 For Dummies
978-0-470-17473-9

Salesforce.com For Dummies, 2nd Edition
978-0-470-04893-1

Word 2007 For Dummies
978-0-470-03658-7

EDUCATION, HISTORY, REFERENCE & TEST PREPARATION

African American History For Dummies
978-0-7645-5469-8

Algebra For Dummies
978-0-7645-5325-7

Algebra Workbook For Dummies
978-0-7645-8467-1

Art History For Dummies
978-0-470-09910-0

ASVAB For Dummies, 2nd Edition
978-0-470-10671-6

British Military History For Dummies
978-0-470-03213-8

Calculus For Dummies
978-0-7645-2498-1

Canadian History For Dummies, 2nd Edition
978-0-470-83656-9

Geometry Workbook For Dummies
978-0-471-79940-5

The SAT I For Dummies, 6th Edition
978-0-7645-7193-0

Series 7 Exam For Dummies
978-0-470-09932-2

World History For Dummies
978-0-7645-5242-7

FOOD, GARDEN, HOBBIES & HOME

Bridge For Dummies, 2nd Edition
978-0-471-92426-5

Coin Collecting For Dummies, 2nd Edition
978-0-470-22275-1

Cooking Basics For Dummies, 3rd Edition
978-0-7645-7206-7

Drawing For Dummies
978-0-7645-5476-6

Etiquette For Dummies, 2nd Edition
978-0-470-10672-3

Gardening Basics For Dummies*†
978-0-470-03749-2

Knitting Patterns For Dummies
978-0-470-04556-5

Living Gluten-Free For Dummies†
978-0-471-77383-2

Painting Do-It-Yourself For Dummies
978-0-470-17533-0

HEALTH, SELF HELP, PARENTING & PETS

Anger Management For Dummies
978-0-470-03715-7

Anxiety & Depression Workbook For Dummies
978-0-7645-9793-0

Dieting For Dummies, 2nd Edition
978-0-7645-4149-0

Dog Training For Dummies, 2nd Edition
978-0-7645-8418-3

Horseback Riding For Dummies
978-0-470-09719-9

Infertility For Dummies†
978-0-470-11518-3

Meditation For Dummies with CD-ROM, 2nd Edition
978-0-471-77774-8

Post-Traumatic Stress Disorder For Dummies
978-0-470-04922-8

Puppies For Dummies, 2nd Edition
978-0-470-03717-1

Thyroid For Dummies, 2nd Edition†
978-0-471-78755-6

Type 1 Diabetes For Dummies*†
978-0-470-17811-9

* Separate Canadian edition also available
† Separate U.K. edition also available

Available wherever books are sold. For more information or to order direct: U.S. customers visit www.dummies.com or call 1-877-762-2974.
U.K. customers visit www.wileyeurope.com or call (0)1243 843291. Canadian customers visit www.wiley.ca or call 1-800-567-4797.

WILEY

INTERNET & DIGITAL MEDIA

AdWords For Dummies
978-0-470-15252-2

Blogging For Dummies, 2nd Edition
978-0-470-23017-6

Digital Photography All-in-One Desk Reference For Dummies, 3rd Edition
978-0-470-03743-0

Digital Photography For Dummies, 5th Edition
978-0-7645-9802-9

Digital SLR Cameras & Photography For Dummies, 2nd Edition
978-0-470-14927-0

eBay Business All-in-One Desk Reference For Dummies
978-0-7645-8438-1

eBay For Dummies, 5th Edition*
978-0-470-04529-9

eBay Listings That Sell For Dummies
978-0-471-78912-3

Facebook For Dummies
978-0-470-26273-3

The Internet For Dummies, 11th Edition
978-0-470-12174-0

Investing Online For Dummies, 5th Edition
978-0-7645-8456-5

iPod & iTunes For Dummies, 5th Edition
978-0-470-17474-6

MySpace For Dummies
978-0-470-09529-4

Podcasting For Dummies
978-0-471-74898-4

Search Engine Optimization For Dummies, 2nd Edition
978-0-471-97998-2

Second Life For Dummies
978-0-470-18025-9

Starting an eBay Business For Dummies, 3rd Edition†
978-0-470-14924-9

GRAPHICS, DESIGN & WEB DEVELOPMENT

Adobe Creative Suite 3 Design Premium All-in-One Desk Reference For Dummies
978-0-470-11724-8

Adobe Web Suite CS3 All-in-One Desk Reference For Dummies
978-0-470-12099-6

AutoCAD 2008 For Dummies
978-0-470-11650-0

Building a Web Site For Dummies, 3rd Edition
978-0-470-14928-7

Creating Web Pages All-in-One Desk Reference For Dummies, 3rd Edition
978-0-470-09629-1

Creating Web Pages For Dummies, 8th Edition
978-0-470-08030-6

Dreamweaver CS3 For Dummies
978-0-470-11490-2

Flash CS3 For Dummies
978-0-470-12100-9

Google SketchUp For Dummies
978-0-470-13744-4

InDesign CS3 For Dummies
978-0-470-11865-8

Photoshop CS3 All-in-One Desk Reference For Dummies
978-0-470-11195-6

Photoshop CS3 For Dummies
978-0-470-11193-2

Photoshop Elements 5 For Dummies
978-0-470-09810-3

SolidWorks For Dummies
978-0-7645-9555-4

Visio 2007 For Dummies
978-0-470-08983-5

Web Design For Dummies, 2nd Edition
978-0-471-78117-2

Web Sites Do-It-Yourself For Dummies
978-0-470-16903-2

Web Stores Do-It-Yourself For Dummies
978-0-470-17443-2

LANGUAGES, RELIGION & SPIRITUALITY

Arabic For Dummies
978-0-471-77270-5

Chinese For Dummies, Audio Set
978-0-470-12766-7

French For Dummies
978-0-7645-5193-2

German For Dummies
978-0-7645-5195-6

Hebrew For Dummies
978-0-7645-5489-6

Ingles Para Dummies
978-0-7645-5427-8

Italian For Dummies, Audio Set
978-0-470-09586-7

Italian Verbs For Dummies
978-0-471-77389-4

Japanese For Dummies
978-0-7645-5429-2

Latin For Dummies
978-0-7645-5431-5

Portuguese For Dummies
978-0-471-78738-9

Russian For Dummies
978-0-471-78001-4

Spanish Phrases For Dummies
978-0-7645-7204-3

Spanish For Dummies
978-0-7645-5194-9

Spanish For Dummies, Audio Set
978-0-470-09585-0

The Bible For Dummies
978-0-7645-5296-0

Catholicism For Dummies
978-0-7645-5391-2

The Historical Jesus For Dummies
978-0-470-16785-4

Islam For Dummies
978-0-7645-5503-9

Spirituality For Dummies, 2nd Edition
978-0-470-19142-2

NETWORKING AND PROGRAMMING

ASP.NET 3.5 For Dummies
978-0-470-19592-5

C# 2008 For Dummies
978-0-470-19109-5

Hacking For Dummies, 2nd Edition
978-0-470-05235-8

Home Networking For Dummies, 4th Edition
978-0-470-11806-1

Java For Dummies, 4th Edition
978-0-470-08716-9

Microsoft® SQL Server™ 2008 All-in-One Desk Reference For Dummies
978-0-470-17954-3

Networking All-in-One Desk Reference For Dummies, 2nd Edition
978-0-7645-9939-2

Networking For Dummies, 8th Edition
978-0-470-05620-2

SharePoint 2007 For Dummies
978-0-470-09941-4

Wireless Home Networking For Dummies, 2nd Edition
978-0-471-74940-0